Everyday Mathematics®

The University of Chicago School Mathematics Project

Skills Link

Cumulative Practice Sets
Student Book

The McGraw·Hill Companies

Photo Credits

Cover—Getty Images, cover, *right*; Frank Lane Picture Agency/CORBIS, cover, *bottom left*; ©Star/sefa/CORBIS, cover, *center*

Photo Collage—Herman Adler Design

www.WrightGroup.com

Printed in the United States of America.

Send all inquiries to:
Wright Group/McGraw-Hill
8787 Orion Place
Columbus, OH 43240

ISBN 978-0-07-622502-6
MHID 0-07-622502-X

6 7 8 9 MAL 15 14 13 12 11 10

The McGraw-Hill Companies

Contents

Practice Sets Correlated to Grade 2 Goals

Content	*Everyday Mathematics* Grade 2 Grade-Level Goals	Grade 2 Practice Sets
Number and Numeration		
Rote counting	**Goal 1.** Count on by 1s, 2s, 5s, 10s, 25s, and 100s past 1,000 and back by 1s from any number less than 1,000 with and without number grids, number lines, and calculators.	1, 2, 4, 6, 7, 8, 11, 15, 26, 49, 50, 51, 59, 60, 88
Place value and notation	**Goal 2.** Read, write, and model with manipulatives whole numbers up to 10,000; identify places in such numbers and the values of the digits in those places; read and write money amounts in dollars-and-cents notation.	6, 8, 10, 12, 16, 19, 20, 21, 22, 24, 27, 32, 34, 36, 51, 74, 75, 77, 79, 82, 87
Meanings and uses of fractions	**Goal 3.** Use manipulatives and drawings to model fractions as equal parts of a region or a collection; describe the models and name the fractions.	58, 59, 60, 61, 62, 63, 65, 66, 73, 89
Number theory	**Goal 4.** Recognize numbers as odd or even.	7, 16, 21, 24, 46, 50, 74
Equivalent names for whole numbers	**Goal 5.** Use tally marks, arrays, and numerical expressions involving addition and subtraction to give equivalent names for whole numbers.	14, 19, 38
Equivalent names for fractions, decimals, and percents	**Goal 6.** Use manipulatives and drawings to model equivalent names for $\frac{1}{2}$.	62, 66
Comparing and ordering numbers	**Goal 7.** Compare and order whole numbers up to 10,000; use area models to compare fractions.	23, 35, 53, 56
Operations and Computation		
Addition and subtraction facts	**Goal 1.** Demonstrate automatically with +/– 0, +/– 1, doubles, and sum-equals-ten facts, and proficiency with all addition and subtraction facts through 10 + 10.	1, 3, 5, 9, 10, 11, 12, 13, 14, 17, 18, 20, 25, 30, 40, 41, 53, 56, 62, 70
Addition and subtraction procedures	**Goal 2.** Use manipulatives, number grids, tally marks, mental arithmetic, paper & pencil, and calculators to solve problems involving the addition and subtraction of 2-digit whole numbers; describe the strategies used; calculate and compare values of coin and bill combinations.	2, 3, 5, 9, 12, 13, 17, 21, 25, 26, 27, 28, 31, 33, 39, 40, 41, 43, 44, 48, 50, 51, 52, 53, 54, 55, 58, 59, 62, 63, 65, 72, 73, 75, 76, 79, 80, 81, 82, 85, 86
Computational estimation	**Goal 3.** Make reasonable estimates for whole number addition and subtraction problems; explain how the estimates were obtained.	44, 75, 81, 82
Models for the operations	**Goal 4.** Identify and describe change, comparison, and parts-and-total situations; use repeated addition, arrays, and skip counting to model multiplication; use equal sharing and equal grouping to model division.	27, 28, 31, 41, 43, 45, 46, 47, 48, 49, 52, 54, 57, 64, 68, 69, 70, 83, 84, 85, 86, 87, 88, 91
Data and Chance		
Data collection and representation	**Goal 1.** Collect and organize data or use given data to create tally charts, tables, bar graphs, and line plots.	23, 42, 92
Data analysis	**Goal 2.** Use graphs to ask and answer simple questions and draw conclusions; find the maximum, minimum, mode, and median of a data set.	23, 42, 56, 57, 92
Qualitative probability	**Goal 3.** Describe events using *certain, likely, unlikely, impossible* and other basic probability terms; explain the choice of language.	9, 11, 37, 67, 74

Content	*Everyday Mathematics* Grade 2 Grade-Level Goals	Grade 2 Practice Sets
Measurement and Reference Frames		
Length, weight, and angles	**Goal 1.** Estimate length with and without tools; measure length to the nearest inch and centimeter; use standard and nonstandard tools to measure and estimate weight.	32, 33, 54, 55, 66, 67, 68, 70, 71, 80
Area, perimeter, volume, and capacity	**Goal 2.** Count unit squares to find the area of rectangles.	64, 69, 76
Units and systems of measurement	**Goal 3.** Describe relationships between days in a week and hours in a day.	67, 89
Money	**Goal 4.** Make exchanges between coins and bills.	3, 20, 25, 28, 31, 67, 76, 78
Temperature	**Goal 5.** Read temperature on both the Fahrenheit and Celsius scales.	29, 30, 72, 91
Time	**Goal 6.** Tell and show time to the nearest five minutes on an analog clock; tell and write time in digital notation.	1, 5, 10, 21, 22, 34, 78
Geometry		
Lines and angles	**Goal 1.** Draw line segments and identify parallel line segments.	35, 37, 49, 73, 78
Plane and solid figures	**Goal 2.** Identify, describe, and model plane and solid figures including circles, triangles, squares, rectangles, hexagons, trapezoids, rhombuses, spheres, cylinders, rectangular prisms, pyramids, cones, and cubes.	35, 37, 38, 83
Transformations and symmetry	**Goal 3.** Create and complete two-dimensional symmetric shapes or designs.	39, 57, 78, 89
Patterns, Functions, and Algebra		
Patterns and functions	**Goal 1.** Extend, describe, and create numeric, visual, and concrete patterns; describe rules for patterns and use them to solve problems; use words and symbols to describe and write rules for functions involving addition and subtraction and use those rules to solve problems.	15, 16, 18, 24, 29, 44, 45, 53, 77, 81, 90
Algebraic notation and solving number sentences	**Goal 2.** Read, write, and explain expressions and number sentences using the symbols $+$, $-$, $=$, $>$, and $<$; solve number sentences involving addition and subtraction; write expressions and number sentences to model number stories.	7, 9, 21, 80, 88
Properties of arithmetic operations	**Goal 3.** Describe the Commutative and Associative Properties of Addition and apply them to mental arithmetic problems.	11, 70, 80, 90

Name Date Time

Grade 1 Review: Number and Numeration

1. Count up by 5s.

   45, 50, 55, 60, 65, 70, ____

2. Count back by 1s.

   34, 33, 32, 31, 30, 21, 22

Circle the larger number.

3. 572 527

4. 109 209

5. Circle the group that has an *odd* number of balls.

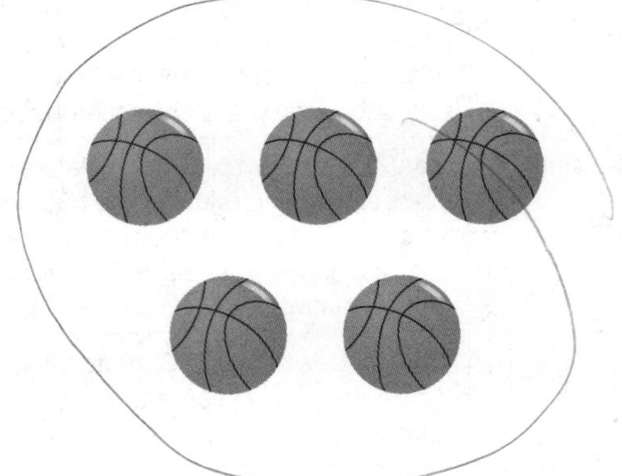

6. How many apples are there? 13 apples

Grade 1 Review: Number and Numeration

7. How many marbles are there? Circle the number.

6 12  27 100

8. Circle $\frac{1}{2}$ of the crayons.

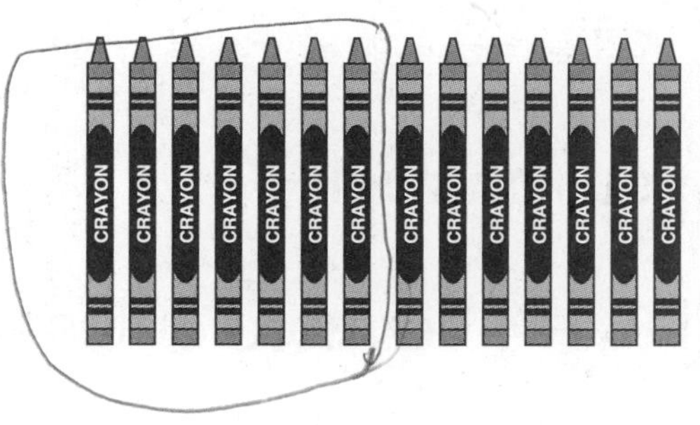

9. Write a number sentence. Tell how many in all.

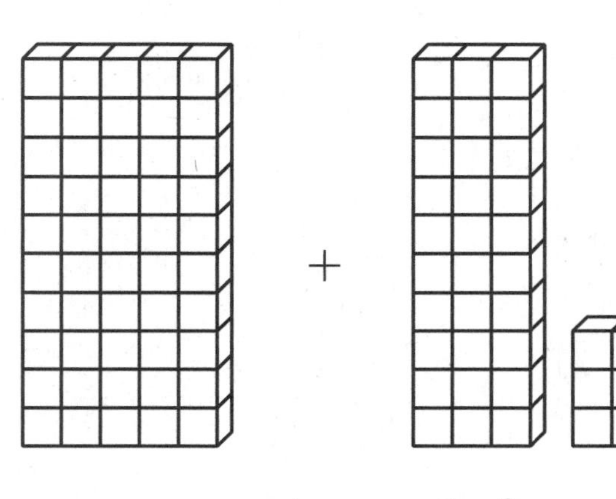

50 + 36 = 86

10. Tell how many.

𝍸 𝍸 𝍸 // 17

11. Use tally marks to show the number 9.

𝍸 ////

Name Date Time

Grade 1 Review: Operations and Computation

Solve.

1. $6 + 6 =$ 12 **2.** $8 - 1 =$ 7 **3.** $5 + 0 =$ 5

Circle your answer.

4. Will $8 + 4$ be more or less than 10?

More Less

5. Will $9 - 2$ be more or less than 5?

More Less

6. Write a number that is more than 87. 1,999

7. Write a number that is less than 72. 62

8. How much money?

56 ¢

9. Sheila has 26 marbles in a bag. Darius puts 9 more marbles in the bag. How many marbles are in the bag now?

35 marbles in all

Name Date Time

Grade 1 Review: Data and Chance

1. Use the table to complete the graph.

Children	Number of Pets
Mario	///
Avi	////
Jose	~~////~~
Jill	//
Todd	///

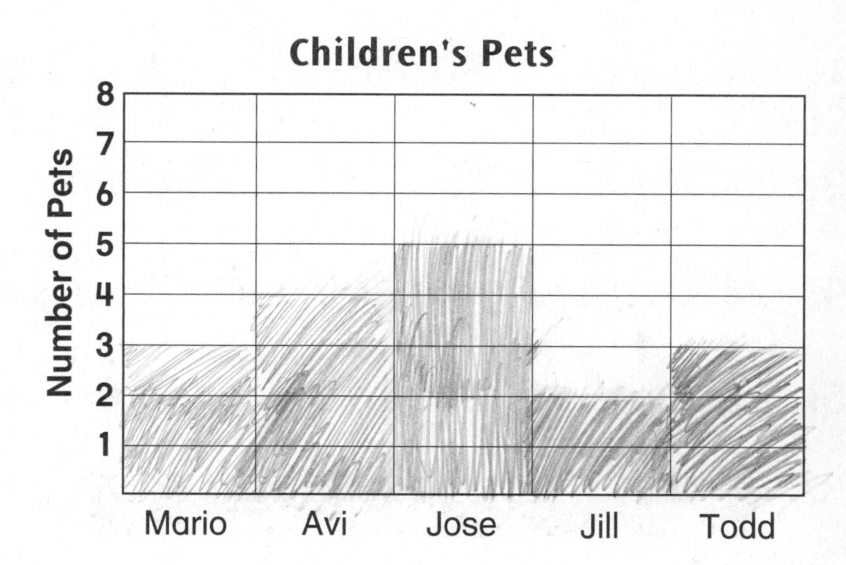

Use the graph to answer the questions.

2. Which child has the most pets? Jose

3. Which child has the fewest pets? Jill

4. Which two children have the same number of pets? Todd and Mario

5. How many pets does Avi have? 4

Use the data below to answer the questions.

52, 23, 81, 71, 23, 94, 60

6. What is the maximum of the data set? 94

7. What is the minimum of the data set? 23

Grade 1 Review: Measurement and Reference Frames

Write <, >, or =.

1. 3 dimes __>__ $0.25

2. 230¢ __<__ $2.30

3. $1.06 __=__ 4 quarters and 6 dimes

4. 130¢ __>__ 3 quarters, 3 dimes, and 4 nickels

5. Estimate the length to the nearest inch.

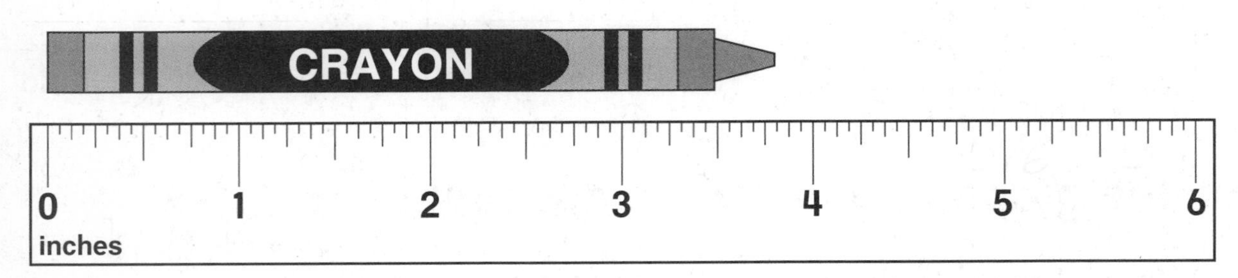

About __4__ inches

6. Write the temperature.

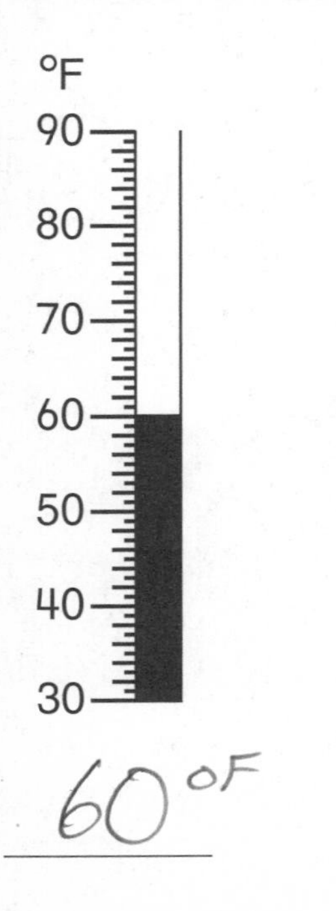

__60 °F__

7. Write the time.

__5:45__

Name Date Time

Grade 1 Review: Geometry

Name the shapes.

triangle **cylinder** **rectangle**
circle **cone** **cube**

1. 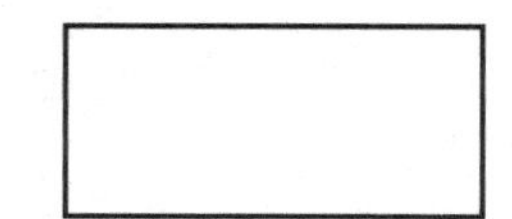

rectangle

2. 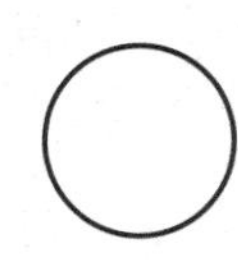

circle

3. 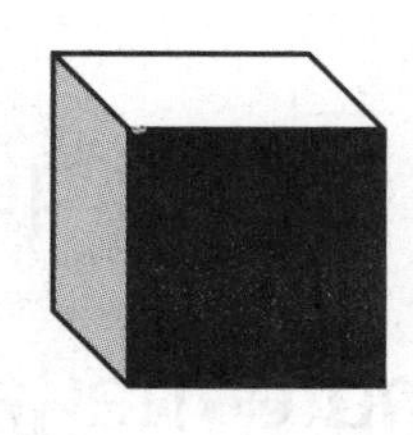

cube

4. 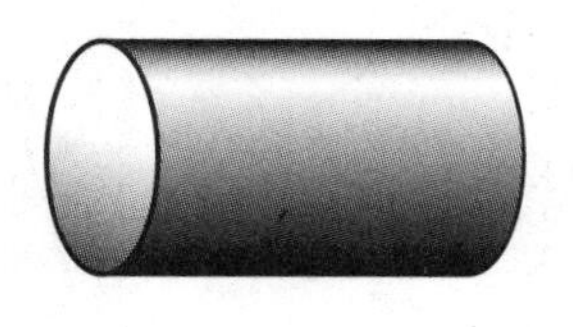

cylinder

5. 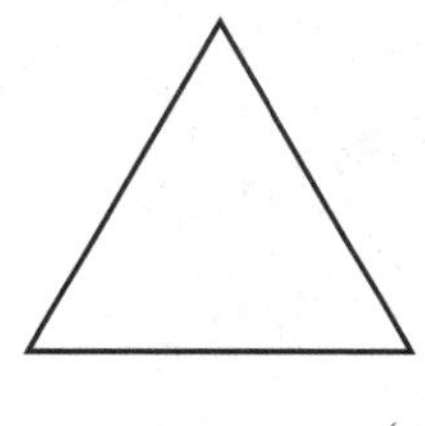

triangle

6. 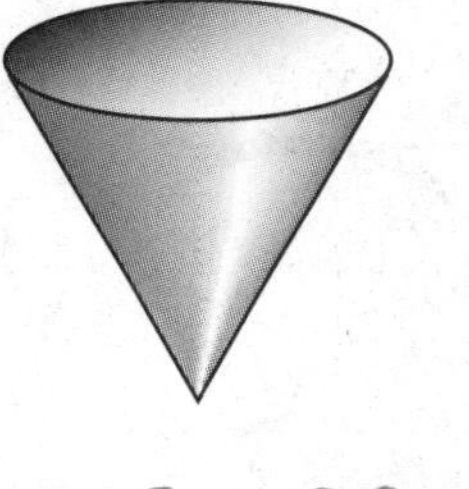

cone

7. How many sides does a square have? 4

8. How many corners does a triangle have? 3

9. How many corners does a circle have? 0

Draw the other half.

10. 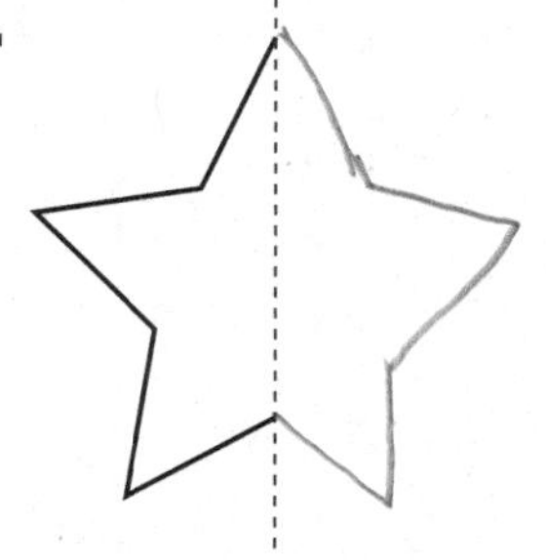

11. 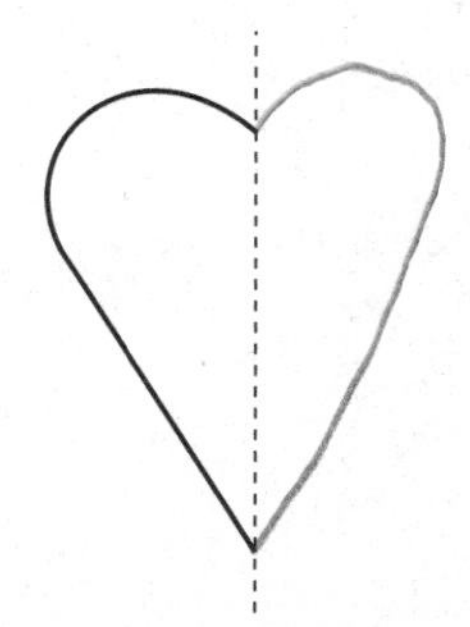

Name Date Time

Grade 1 Review: Patterns, Functions, and Algebra

Fill in each number-grid piece.

1.

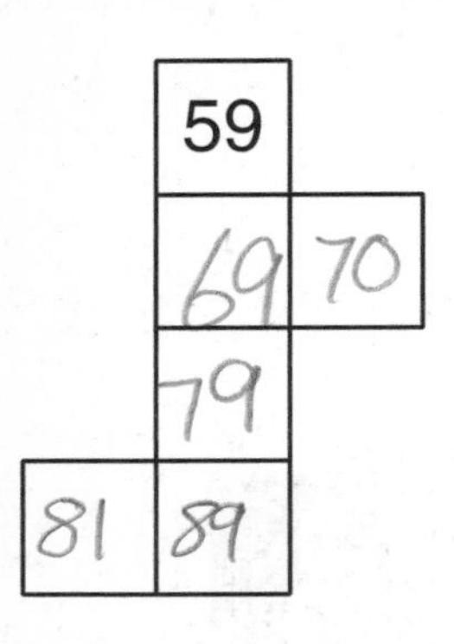

2.

98	99	100
108	109	
	129	130
	139	

Fill in the rule box. Then complete the frames.

3.

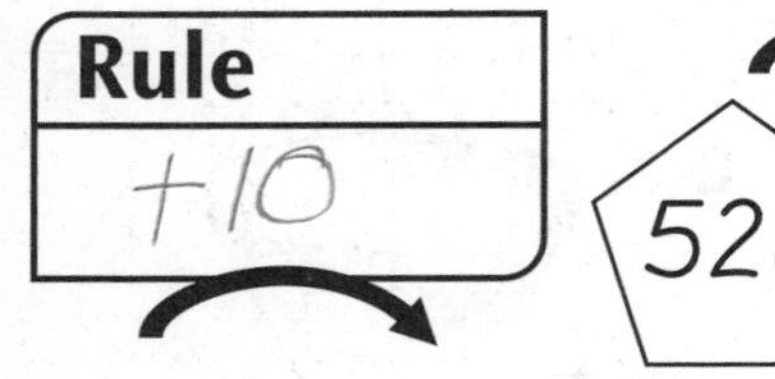

546

576 586

4.

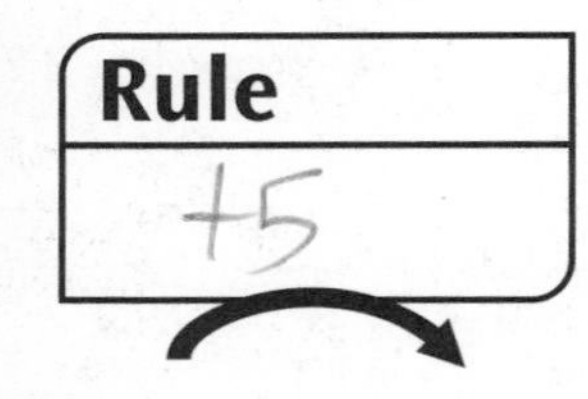

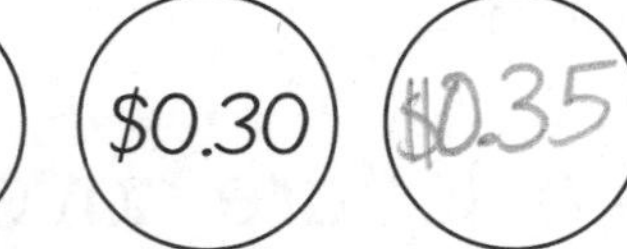

$0.35

$0.45 $0.50

Write the turn-around fact.

5. 6 + 2 = 8

2 + 6 = 8

6. 7 + 8 = 15

8 + 7 = 15

7. 5 + 7 = 12

7 + 5 = 12

8. 3 + 8 = 11

8 + 3 = 11

Name Date Time

Practice Set 1

Add.

1. 12

2. 15

Fill in the missing numbers.

3. 333, 334, 335, 336, 337, 338, 339, 340, 341, 342

4. 718, 719, 720, 721, 722, 723, 724, 724, 725, 726

Draw the hands to show the time.

5.

7:00

6.

10:30

7.

3:15

Name Date Time

Practice Set 2

How much money? Write the amount.

1\.

$ 20.00

2\.

$ 152.00

3\.

$ 15.41

Fill in the missing numbers on the number line.

4\.

97 98 99 100 101 102 103 104

5\.

2,999 3,000 3,001 3,002 3,003 3,004 3,005

Fill in the missing numbers.

6\. 5, 10, 15, 20, 25

7\. 302, 304, 306, 308

Use with or after Lesson 1•5.

Name Date Time

Practice Set 3

Add. Remember to practice and memorize your addition facts.

1. $3 + 3 =$ 6
2. $2 + 1 =$ 3
3. $0 + 6 =$ 6
4. $4 + 2 =$ 6
5. $5 + 3 =$ 8

6. 5 $= 3 + 2$
7. $1 + 4 =$ 5
8. 4 $= 2 + 2$
9. $3 + 0 =$ 3
10. 9 $= 4 + 5$
11. $2 + 5 =$ 7

Write the total amount.

Example     = 61 ¢

12.     = 27 ¢

13.       = 39 ¢

14.        = 47 ¢

How many?

15. 1  = 2 

16. 1 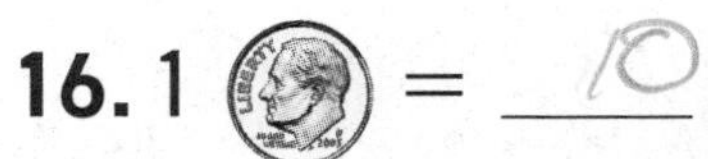  = 10 

Name Date Time

Practice Set 4

501	502	503	504	505	506	507	508	509	510
511	512	513	514	515	516	517	518	519	520
521	522	523	524	525	526	527	528	529	530
531	532	533	534	535	536	537	538	539	540
541	542	543	544	545	546	547	548	549	550
551	552	553	554	555	556	557	558	559	560
561	562	563	564	565	566	567	568	569	570
571	572	573	574	575	576	577	578	579	580
581	582	583	584	585	586	587	588	589	590
591	592	593	594	595	596	597	598	599	600
601	602	603	604	605	606	607	608	609	610
611	612	613	614	615	616	617	618	619	620

1. Fill in all of the missing numbers on the grid.

2. Now start at 501 and count by 2s.
   Put an **X** over each number as you count.

3. Now start at 510 and count by 10s.
   Color each number **blue** as you count.

4. Color the number that is 1 more than 513 **red.**

5. Color the number that is 5 more than 601 **green.**

6. Color the number that is 10 less than 564 **yellow.**

Name Date Time

Practice Set 5

Finish these names for 11.

1. 5 + ____ = 11

2. 4 + ____ = 11

3. 3 + ____ = 11

4. ____ + 2 = 11

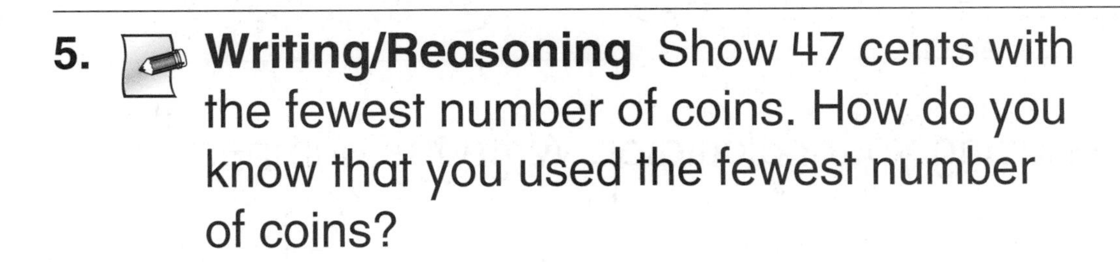

5. **Writing/Reasoning** Show 47 cents with the fewest number of coins. How do you know that you used the fewest number of coins?

____

____

Draw the hands to show the time.

6. 

7:30

7. 

12:45

8. 

5:15

Add. Remember to practice and memorize your addition facts.

9. 6 + 4 = ____

10. 3 + 5 = ____

11. 9 + 2 = ____

Name Date Time

Practice Set 6

Use your calculator to count.

To count by 2s	To count by 5s	To count by 10s
Press (2) and (+). Then press (=) over and over.	Press (5) and (+). Then press (=) over and over.	Press (1)(0) and (+). Then press (=) over and over.

1. Count by 2s using your calculator. Write the numbers.

   2, 4, 6, ___, ___, ___, ___, ___, ___

2. Count by 5s using your calculator. Write the numbers.

   5, 10, 15, ___, ___, ___, ___, ___, ___

3. Count by 10s using your calculator. Write the numbers.

   10, 20, ___, ___, ___, ___, ___, ___, ___

Match.

4. seven dollars and twenty-seven cents — $2.17
5. seventy-two cents — $7.02
6. two dollars and seventeen cents — $0.72
7. seven dollars and two cents — $7.27
8. two dollars and seventy cents — $2.70

Name Date Time

Practice Set 7

Write <, >, or =.

< means *is less than*
> means *is greater than*
= means *is the same as*

1. 49 ____ 94

2. 9 − 4 ____ 4

3. 347 ____ 374

4. 10 ____ 3 + 6

5. 921 ____ 919

6. 3 + 3 ____ 10 − 5

Count by 5s.

7. 5, 10, ____, ____, ____, 30, ____, ____, 45

8. 65, ____, ____, ____, ____, ____, ____, 100, ____

9. 120, 125, ____, ____, ____, ____, ____, ____, 160

10. Fill in the missing numbers.

351	352	353	354						360
	362			365					
371		373				377	378		380
			384					389	

11. Circle the *even* numbers.

420 103 329 525 218

636 774 481 192 847

Name _______________ Date _______ Time _______

Practice Set 8

Write the number.

1. 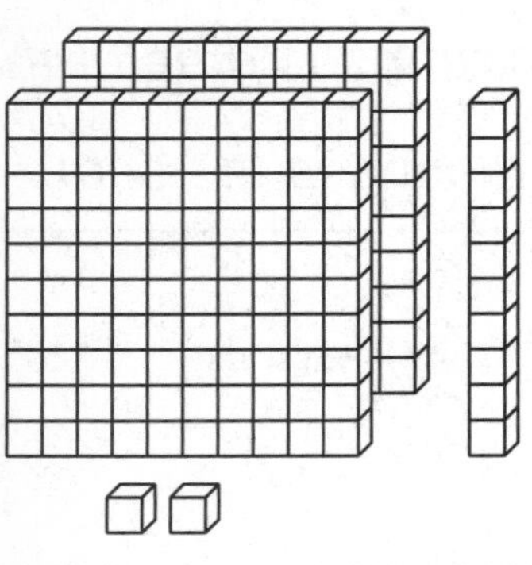

____

2. 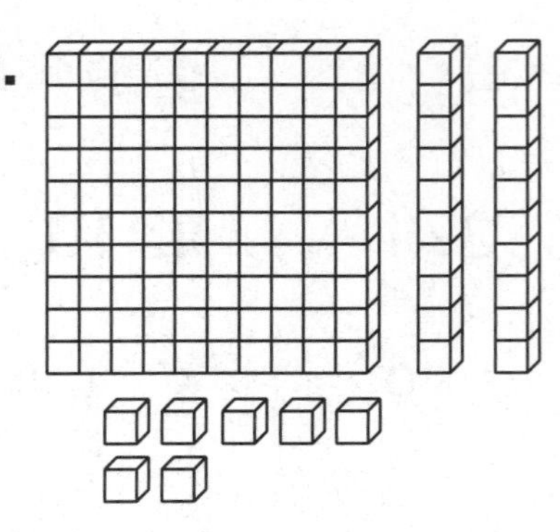

____

Fill in the blanks.

Example 376 = 3 hundreds 7 tens 6 ones

3. 923 = ____ hundreds ____ tens ____ ones

4. 485 = ____ hundreds ____ tens ____ ones

5. 709 = ____ hundreds ____ tens ____ ones

Count by 10s.

6. 100, 110, ____, ____, ____, 150, ____, ____, 180

7. 220, ____, 240, ____, ____, ____, 280, ____, 300

8. **Writing/Reasoning** Which has more tens, 626 or 247? Explain your answer.

____

____

Use with or after Lesson 1•12.

Name Date Time

Practice Set 9

1. **Writing/Reasoning** Write an addition story that asks a question.

____

____

____

Answer the question: ____________________

Write a number model: ______ + ______ = ______

2. Show $0.75 two ways. Use Ⓠs, Ⓓs, and Ⓝs.

Circle the best answer.

3. How likely is it that it will snow today where you live?
   certain likely unlikely impossible

4. How likely is it that the sun will rise tomorrow?
   certain likely unlikely impossible

Name Date Time

Practice Set 10

Add. Remember to practice and memorize your addition facts.

1. $4 + 4 =$ ____
2. $7 + 7 =$ ____
3. $8 + 8 =$ ____
4. $9 + 9$
5. $8 + 9$
6. $7 + 6$
7. $6 + 5$

Circle the time shown on the clock.

8. 

7:15 6:30

9. 

6:00 12:30

Write the number.

Example three hundred fifty-two 352

10. eight hundred twenty-nine ______
11. two hundred seventeen ______
12. seven hundred nine ______
13. ninety-one ______

Use with or after Lesson 2•3.

Name Date Time

Practice Set 11

Write the turn-around addition facts for each domino.

Example

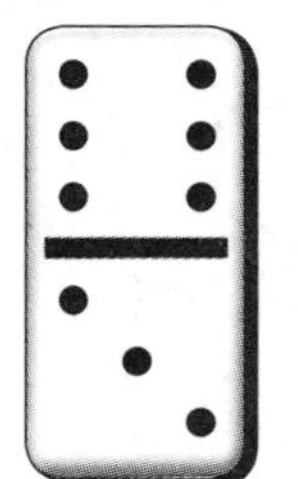

6 + 3 = 9

3 + 6 = 9

1. 

______

______

2. 

______

______

3. 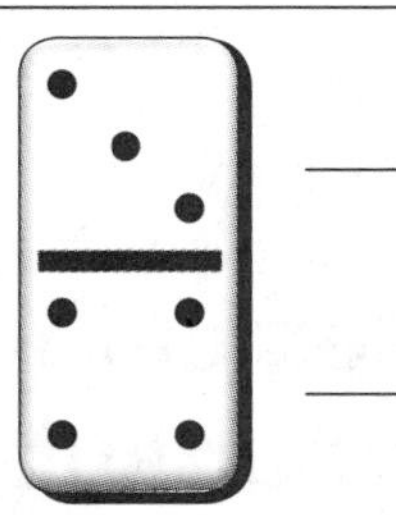

______

______

Count back by 10s.

4. 200, 190, 180, ____, ____, ____, ____, ____, 120

5. 350, 340, ____, ____, ____, ____, 290, ____, ____

6. 620, 610, ____, ____, 580, ____, ____, ____, ____

Look at the spinner.

Circle the best answer.

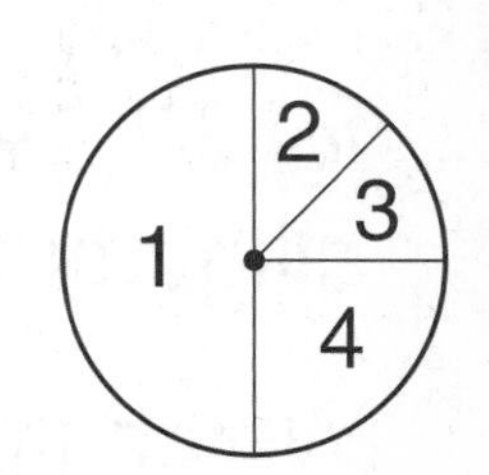

7. How likely is it that the spinner will land on 1?
certain likely unlikely impossible

8. How likely is it that the spinner will land on 6?
certain likely unlikely impossible

Name Date Time

Practice Set 12

1. Write 4 number sentences to match the picture.

_____ + _____ = _____

_____ + _____ = _____

_____ − _____ = _____

_____ − _____ = _____

Subtract. Remember to practice and memorize your subtraction facts.

2. 4 − 0 = _____ **3.** 7 − 1 = _____ **4.** 8 − 0 = _____

5. 5 − 1 = _____ **6.** 3 − 2 = _____ **7.** 10 − 1 = _____

8. $\begin{array}{r} 12 \\ -\ 1 \\ \hline \end{array}$ **9.** $\begin{array}{r} 10 \\ -\ 9 \\ \hline \end{array}$ **10.** $\begin{array}{r} 7 \\ -\ 0 \\ \hline \end{array}$ **11.** $\begin{array}{r} 6 \\ -\ 1 \\ \hline \end{array}$

12. A teddy bear is on sale for $4.99. A stuffed alligator costs $5.19.

Which item costs more?

____

13. A bunch of bananas costs $3.45. A bag of apples costs $7.00.

Which item costs less?

____

Name Date Time

Practice Set 13

Write the fact family for the Fact Triangle.

Example

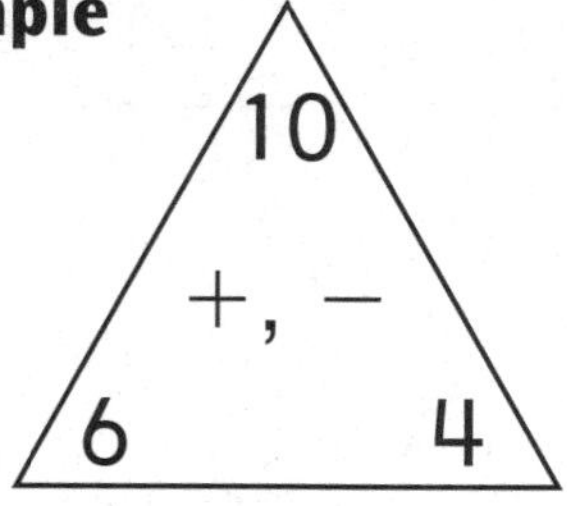

6 + 4 = 10
4 + 6 = 10
10 – 6 = 4
10 – 4 = 6

1.

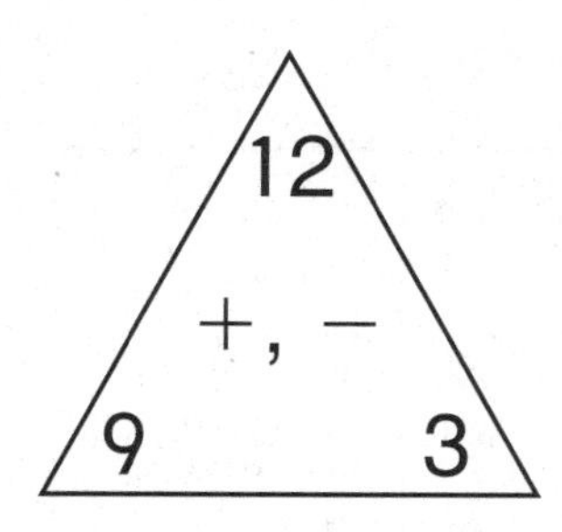

____ + ____ = ____
____ + ____ = ____
____ – ____ = ____
____ – ____ = ____

2. Show 57¢ two different ways.

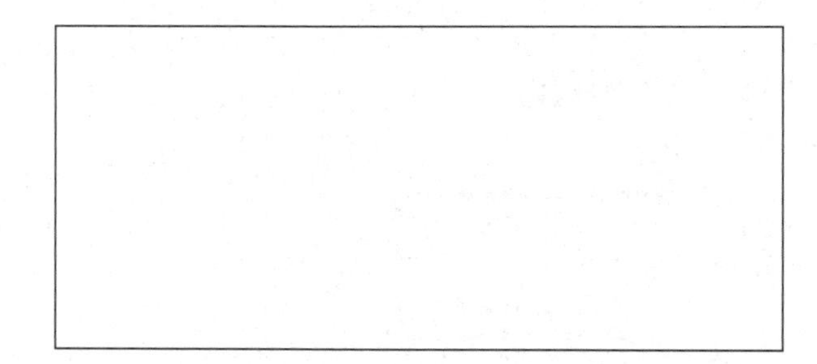
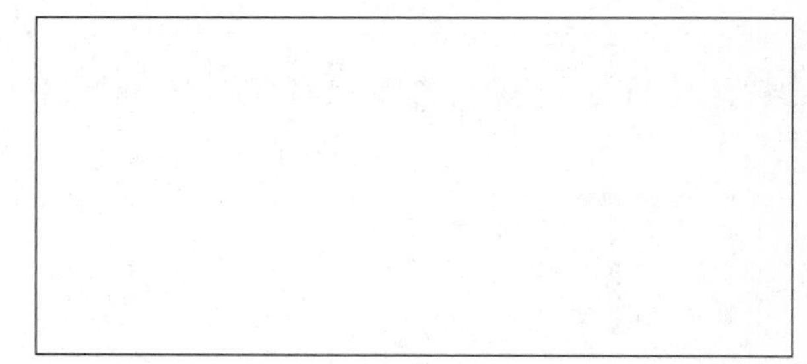

3. Show 81¢ two different ways.

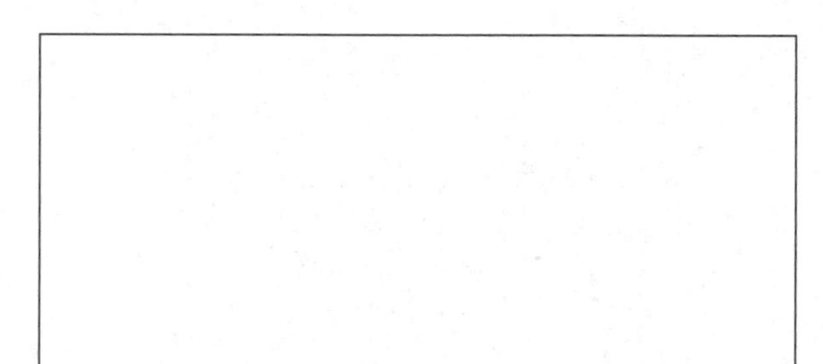
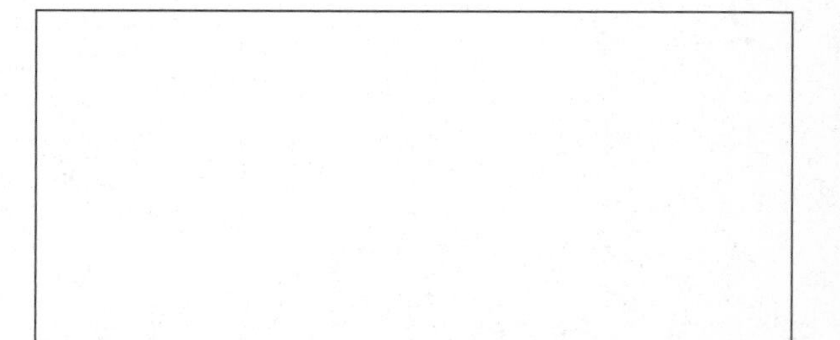

Name Date Time

Practice Set 14

Write 5 names for the number.

Example

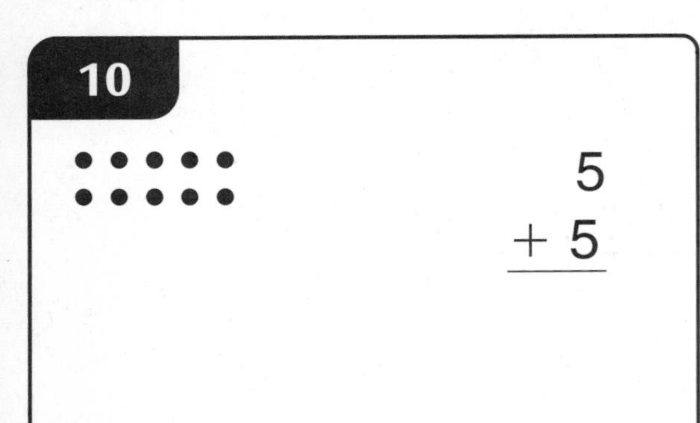

8 + 2 11 − 1

1.

16

Complete the fact family for each domino.

2. 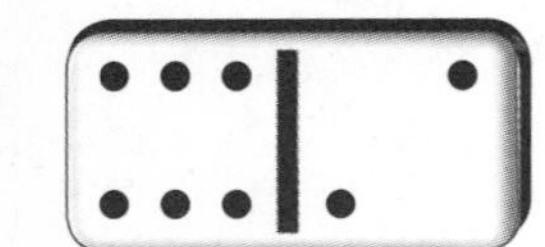

6 + 2 = ______

2 + 6 = ______

8 − 6 = ______

8 − 2 = ______

3. 

3 + 4 = ______

4 + 3 = ______

7 − 3 = ______

7 − 4 = ______

Name Date Time

Practice Set 15

Write the rule.

1. 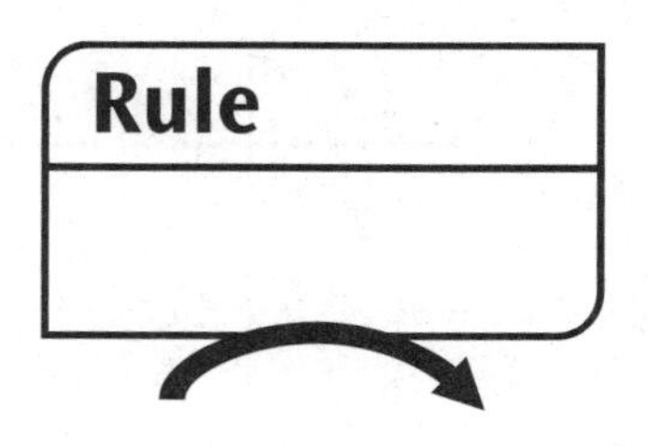

82  84, 86, 88, 90

2. 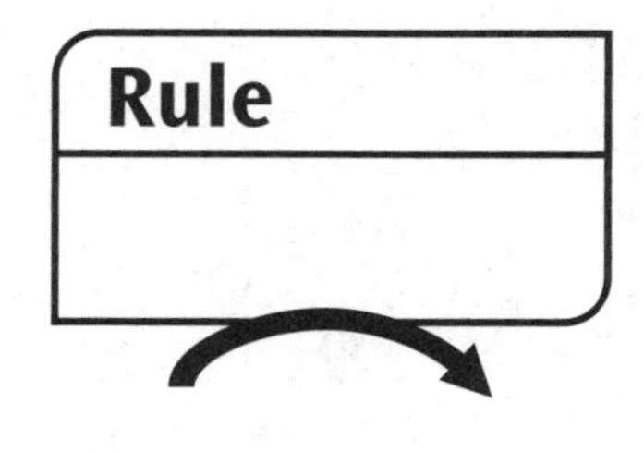

67  57, 47, 37, 27

Fill in the frames. Then write the rule.

3. 

8 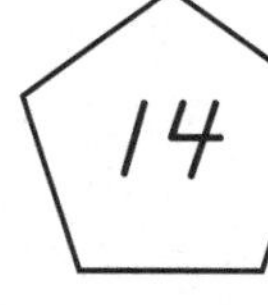 14, 20, 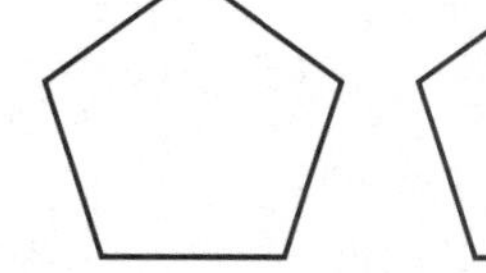

4. 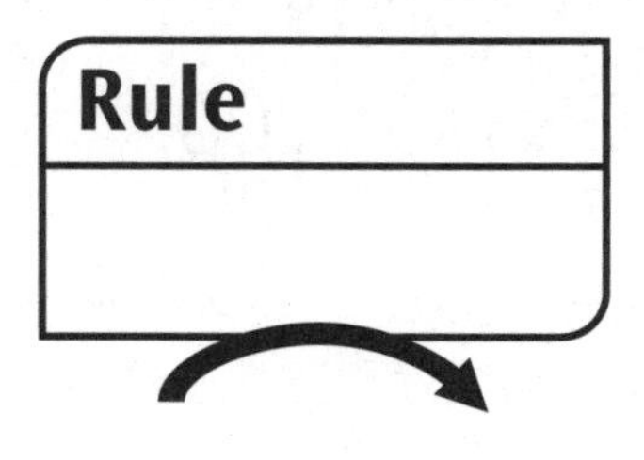

25, 22, ____, ____, 13

Count back by 10s.

5. 450, 440, ____, ____, ____, ____, ____, ____, ____

6. 92, 82, ____, ____, ____, ____, ____, ____, 12

7. 137, ____, ____, ____, 97, ____, ____, ____, ____

Name ____________________ Date ____________ Time ____________

Practice Set 16

Complete the tables.

1.

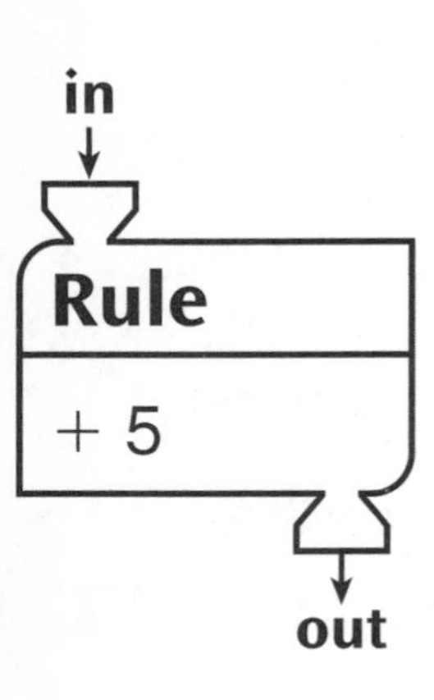

in	out
6	11
8	13
9	
3	
7	

2.

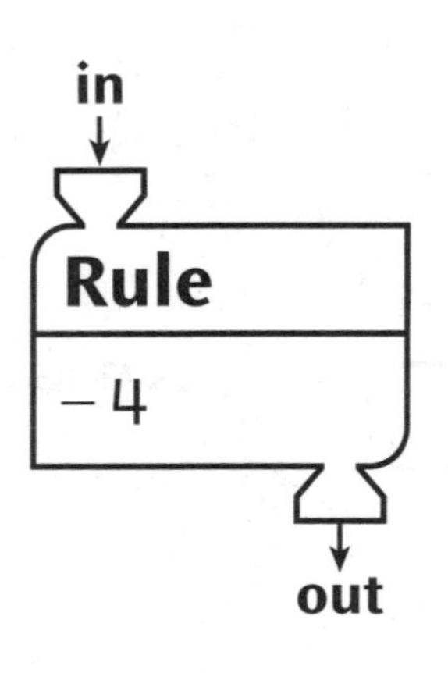

in	out
10	
7	
8	
12	
9	

Write *even* or *odd* for each number.

3. 7 ________ **4.** 19 ________ **5.** 6 ________

6. 42 ________ **7.** 67 ________ **8.** 94 ________

9. Write three *even* numbers between 50 and 100.

________ ________ ________

10. Write three *odd* numbers between 50 and 100.

________ ________ ________

11. Write a 3-digit number that has ...
an *even* number in the hundreds place,
an *odd* number in the tens place, and
an *odd* number in the ones place.

____

Use with or after Lesson 2•11.

Name ____________________ Date ________ Time ________

Practice Set 17

Subtract.

1. $9 - 7 =$ ____ **2.** $11 - 6 =$ ____ **3.** $15 - 5 =$ ____

4. $12 - 7 =$ ____ **5.** $10 - 8 =$ ____ **6.** $13 - 9 =$ ____

7. $16 - 9$ **8.** $14 - 9$ **9.** $13 - 6$ **10.** $11 - 5$

11. $16 - 10$ **12.** $24 - 10$ **13.** $31 - 10$ **14.** $17 - 10$

Write the missing number in the triangle. Then write the fact family.

15.

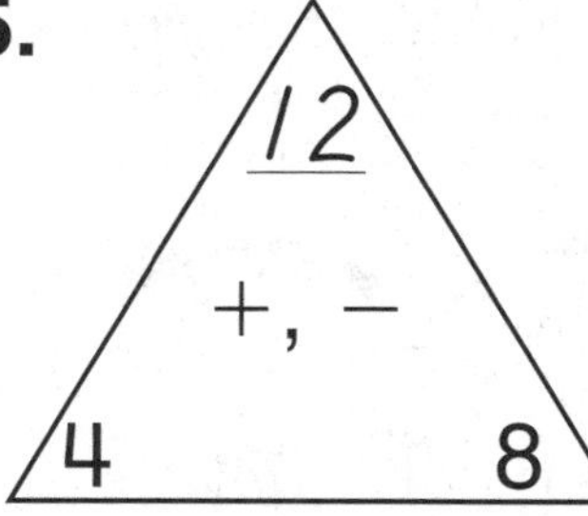

____
____
____
____

16.

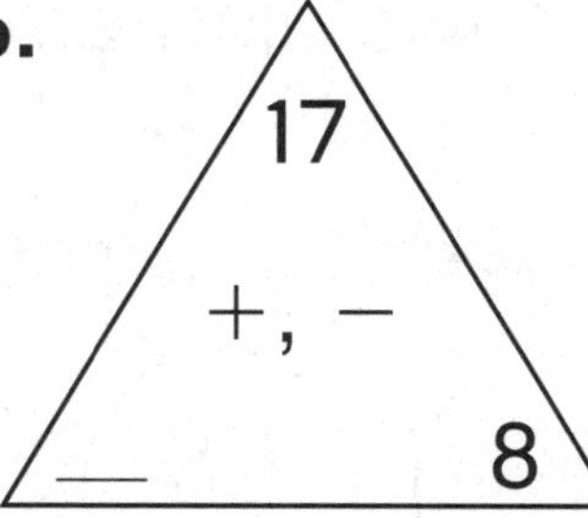

____
____
____
____

17.

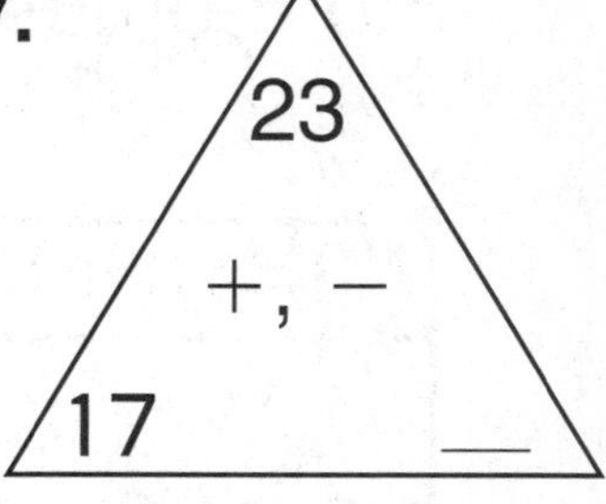

____
____
____
____

Name Date Time

Practice Set 18

Subtract. Use the − 9 and − 8 shortcuts.

Example	To find 21 − 9, think 21 − 10 + 1. To find 21 − 8, think 21 − 10 + 2.

1. $\begin{array}{r} 14 \\ -9 \\ \hline \end{array}$ **2.** $\begin{array}{r} 19 \\ -8 \\ \hline \end{array}$ **3.** $\begin{array}{r} 11 \\ -8 \\ \hline \end{array}$ **4.** $\begin{array}{r} 16 \\ -9 \\ \hline \end{array}$

5. $\begin{array}{r} 15 \\ -8 \\ \hline \end{array}$ **6.** $\begin{array}{r} 12 \\ -9 \\ \hline \end{array}$ **7.** $\begin{array}{r} 18 \\ -8 \\ \hline \end{array}$ **8.** $\begin{array}{r} 20 \\ -9 \\ \hline \end{array}$

9. 17 − 8 = _____ **10.** _____ = 14 − 9 **11.** _____ = 11 − 9

12. 13 − 8 = _____ **13.** 15 − 9 = _____ **14.** 16 − 8 = _____

15. **Writing/Reasoning** Describe the number pattern in the Frames-and-Arrows diagram. Then write the rule.

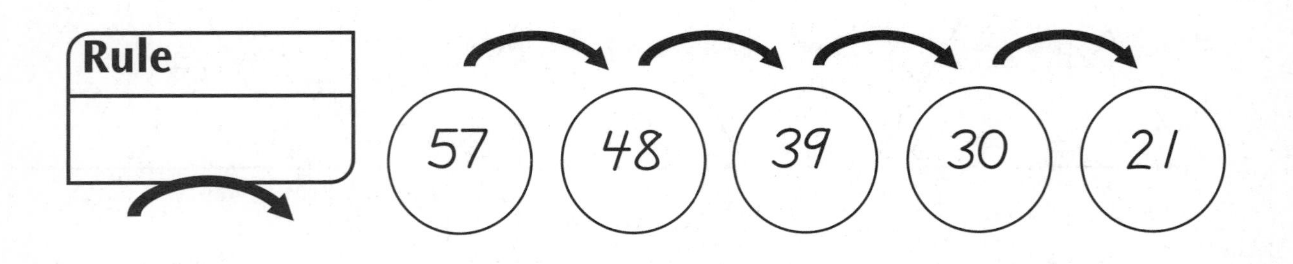

____

____

____

Name Date Time

Practice Set 19

Write the 3-digit number that has ...

Example
8 in the hundreds place
4 in the tens place
6 in the ones place
846

1. 5 in the hundreds place
2 in the tens place
0 in the ones place
______

2. 4 in the hundreds place
9 in the tens place
3 in the ones place
______

3. 2 in the hundreds place
7 in the tens place
8 in the ones place
______

Label the box. Then add four more names to the box.

Example

9

10 − 1

$\begin{array}{r} 8 \\ +\ 1 \\ \hline \end{array}$

HHH ////

4 + 5

6 + 3

$\begin{array}{r} 8 \\ +\ 1 \\ \hline \end{array}$

$\begin{array}{r} 9 \\ +\ 0 \\ \hline \end{array}$

4.

6 + 6

13 − 1

3 + 9

Practice Set 20

How many?

1. 3  = _____ 

2. _____  = 2

3. _____  = 1 

4. 2  = _____ 

5. 1  = _____  and _____ 

Use the three digits. Write the least and the greatest numbers.

Digits	Least Number	Greatest Number
6. 9, 2, 7		
7. 1, 8, 5		
8. 3, 4, 6		
9. 7, 2, 7		

Subtract. Remember to practice and memorize your subtraction facts.

10. _____ = 6 − 3 **11.** 7 − 5 = _____ **12.** _____ = 8 − 2

13. 9 − 4 = _____ **14.** _____ = 4 − 2 **15.** 10 − 3 = _____

16. $\begin{array}{r} 3 \\ -\,2 \\ \hline \end{array}$ **17.** $\begin{array}{r} 8 \\ -\,5 \\ \hline \end{array}$ **18.** $\begin{array}{r} 9 \\ -\,3 \\ \hline \end{array}$ **19.** $\begin{array}{r} 7 \\ -\,0 \\ \hline \end{array}$ **20.** $\begin{array}{r} 6 \\ -\,1 \\ \hline \end{array}$

Name Date Time

Practice Set 21

Tell the time.

1. 

just before ________ o'clock

2. 

about ________ o'clock

How much money?

3.    

____

4.        

____

Write <, >, or =.

5. 2 quarters _____ 5 dimes

6. 5 dimes _____ 7 nickels

7. 6 nickels _____ 35 cents

8. 4 quarters _____ $1.10

Write *odd* or *even*.

9. 27 ________

10. 14 ________

11. 92 ________

Name Date Time

Practice Set 22

Write the number.

1.

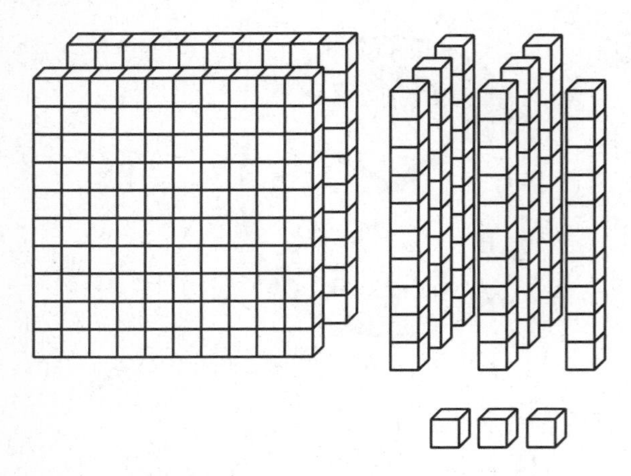

____

2.

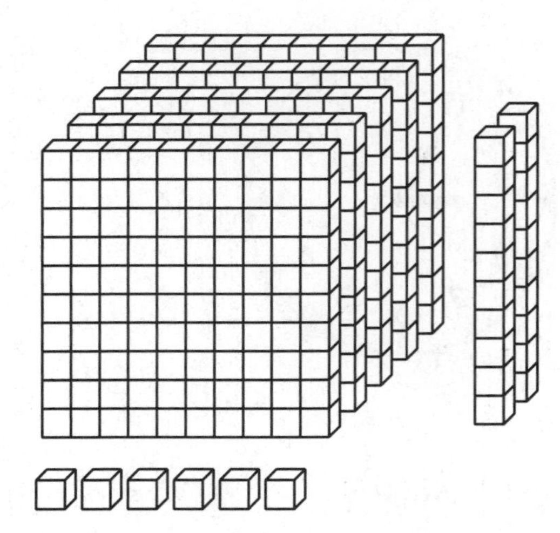

____

How much money?

3.

$ ________________

4.

$ ________________

Circle the time that matches the clock face.

5.

6:10 1:30

6.

9:45 10:45

Use with or after Lesson 3•4.

Name Date Time

Practice Set 23

Each child voted for a favorite sport.
The results are shown in the tally chart.

Favorite Sports	
Football	///
Basketball	////
Baseball	𝍸 //
Soccer	𝍸 ////

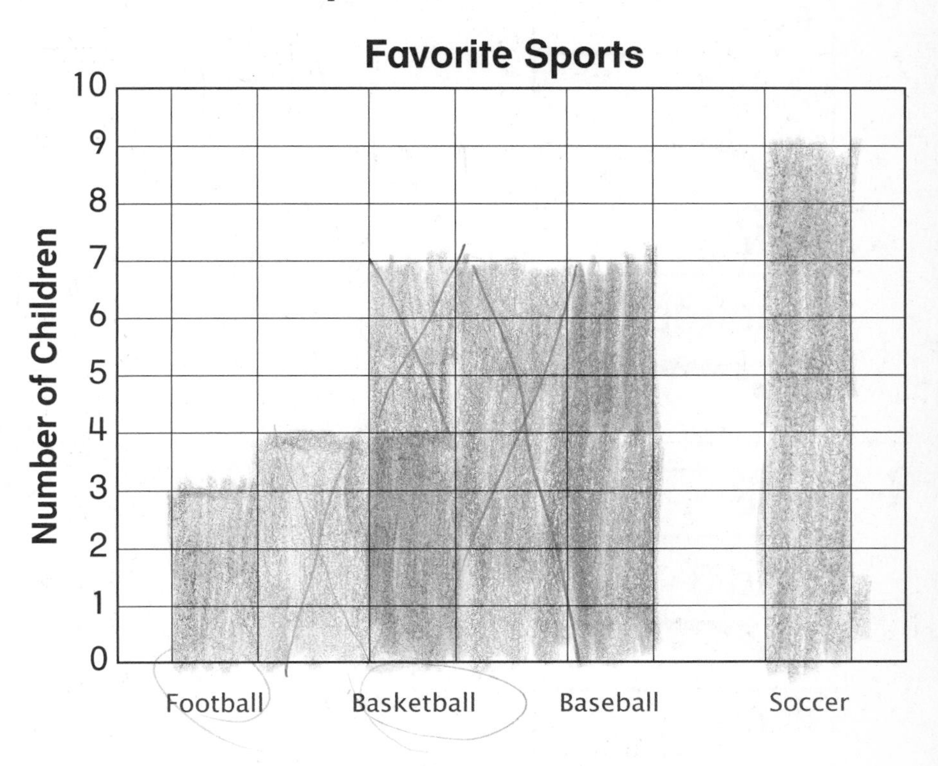

1. Use the tally chart to fill in the bar graph.

2. **Writing/Reasoning** How many more children chose soccer than chose baseball? Explain how you found your answer.

2, because Soccer has 9 and basefall has 7.
9 - 7 = 2

Place these numbers in order.
Circle the middle number.

3. 13, 87, 76, 10, 72

10, 13, 72, 76, 87

4. 1,001; 999; 1,000; 1,011; 997

997; 999; 1,000; 1,001; 1,011

Name Date Time

Practice Set 24

Fill in the frames.

1.

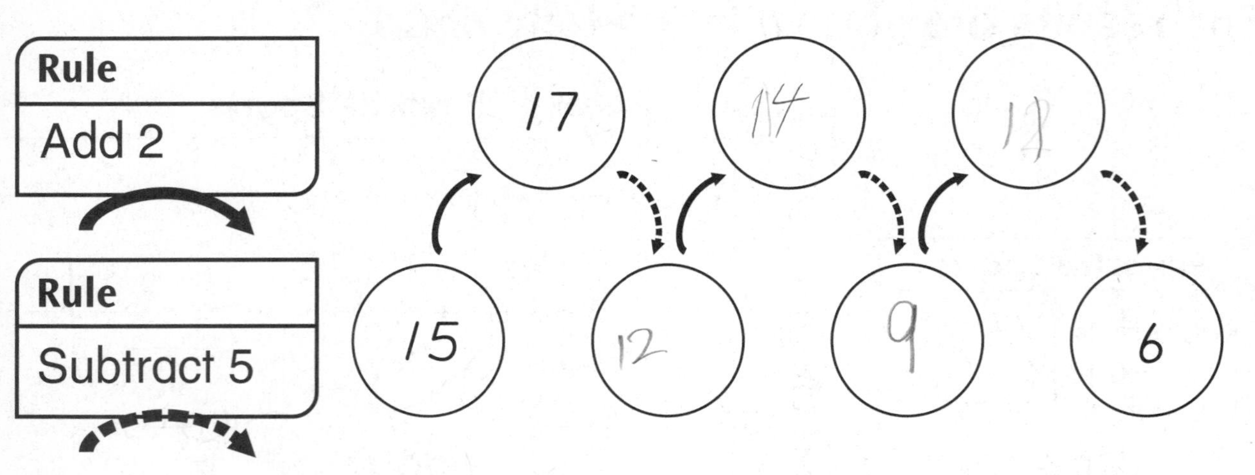

2.

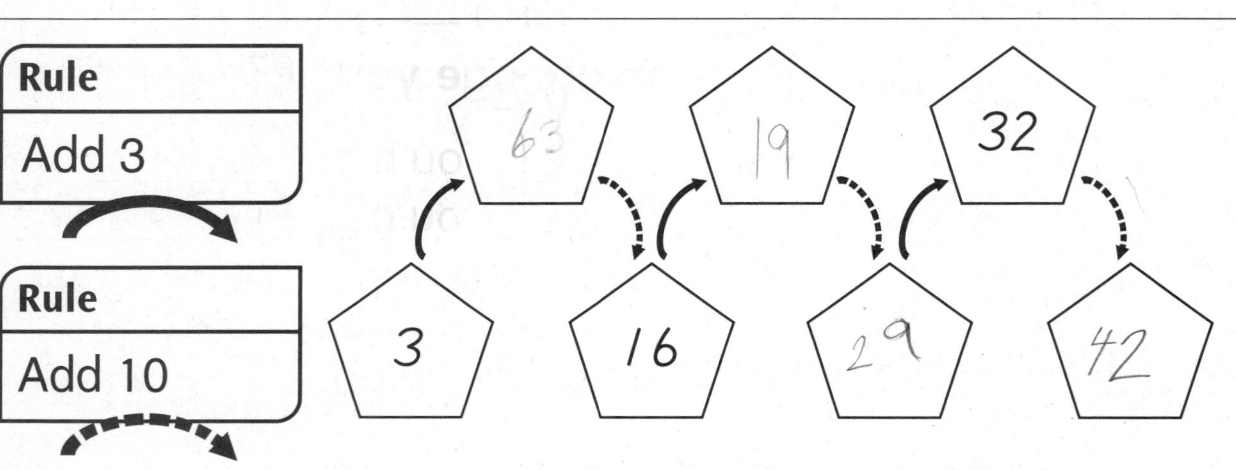

Write the number.

3. six hundred twenty-nine 629
4. two hundred ninety-six 296
5. nine hundred sixteen 916

6. Write 2 *even* numbers. 10 20
7. Write 2 *odd* numbers. 21 31

Name Date Time

Practice Set 25

Count up to find how much change you will get.

1. You buy a box of raisins. You give the clerk 2 dimes. How much change do you get?
   20¢−18¢ =2¢

2. You buy a yogurt. You give the clerk 2 quarters. How much change do you get?
   50¢ - 40¢=10

3. You buy a banana. You give the clerk 1 dime. How much change do you get?
   10¢ −8¢=2¢

4. You buy a bag of pretzels. You give the clerk 2 dimes. How much change do you get?

   20-20=0

Add or subtract. Remember to practice and memorize your basic facts.

5. $4 + 4 =$ 8
6. $8 + 2 =$ 10
7. $9 + 6 =$ 15
8. $10 - 10 =$ 0
9. $12 - 0 =$ 12
10. $7 - 5 =$ 2

Name Date Time

Practice Set 26

Show the price of each item using the fewest coins possible. Draw Ⓠs, Ⓓs, Ⓝs, or Ⓟs.

1.

2.

3.

Q D N P P P

4. **Writing/Reasoning** You have two quarters. Can you buy a box of nuts? Explain your answer.

No, because the nut is 62 ¢ and I have only 50 ¢, I wauld need 12 ¢ more cents.

Count by 3s.

5. 3, 6, 9, 12, 15, 18, 21, 24, 27, 30

6. 20, 23, 26, 29, 32, 35, 38, 41, 44, 47

7. 52, 55, 58, 61, 64, 67, 70, 73, 73, 79

8. 100, 103, 106, 109, 112, 115, 118, 121, 124, 127

Name Date Time

Practice Set 27

Finish each diagram. Then write a number model.

Example

Start	Change	End
21	+14	35

21 + 14 = 35

1.

Start	Change	End
31	+25	56

31 + 25 = 56

2.

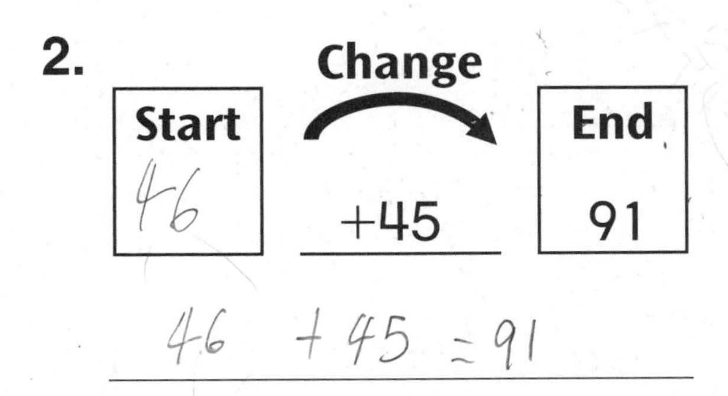

Start	Change	End
46	+45	91

46 + 45 = 91

3.

Start	Change	End
66	+27	93

66 + 27 = 93

Use the three digits. Write the smallest and the largest numbers.

Example

3, 9, 1 smallest 139 largest 931

4. 4, 7, 6 smallest 467 largest 764

5. 2, 6, 3 smallest 236 largest 632

6. 8, 5, 9 smallest 589 largest 985

7. 4, 3, 6 smallest 436 largest 643

8. 5, 2, 8 smallest 258 largest 852

Name Date Time

Practice Set 28

Finish each diagram. Then write a number model.

Example

Total 15	
Part 7	Part 8

7 + 8 = 15

1.

Total 32	
Part 21	Part 11

21 + 11 = 32

2.

Total 61	
Part 33	Part 34

33 + 34 = 61

3.

Total 21		
Part 7	Part 9	Part 5

7 + 9 + 5 = 21

What's your change?

4. You buy a package of crackers for 37 cents. You give the clerk 2 quarters.

Your change: 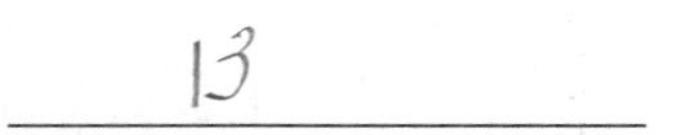 13

5. Nuts cost 62 cents. You pay with 3 quarters.

Your change:  13

Name Date Time

Practice Set 29

Color the thermometer to show each temperature.

Example

47°F

1. 59°F

2. 55°F

3. 62°F

Write the rule. Then fill in the table.

Example

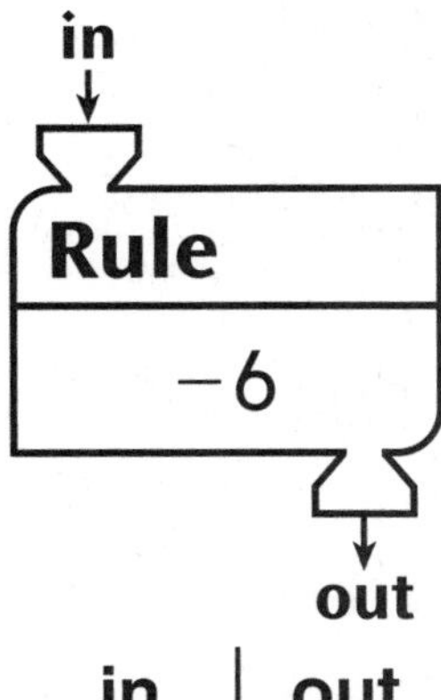

in	out
6	0
9	3
12	6
18	12

4.

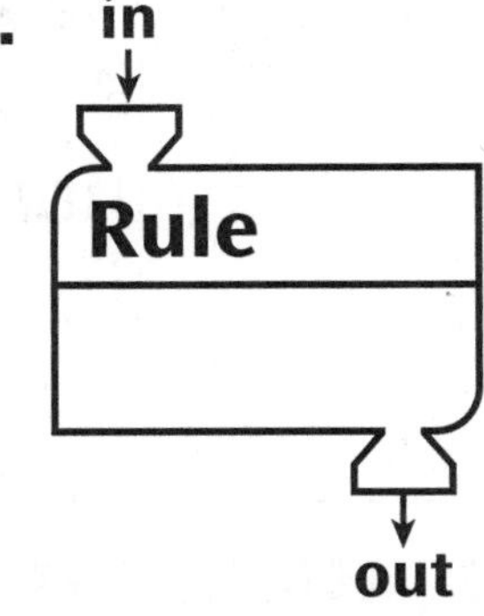

in	out
1	8
0	7
2	
10	

5.

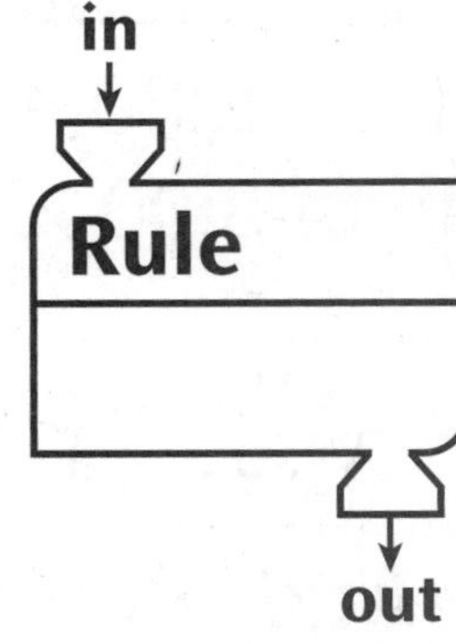

in	out
2	4
8	
3	6
5	

Name Date Time

Practice Set 30

Record the temperature.

Example
8:00 A.M.

°F 100 90 80 70 60 50 40 30 20

30 °F

1.
12:30 P.M.

°F 100 90 80 70 60 50 40 30 20

_____ °F

2.
5:00 P.M.

°F 100 90 80 70 60 50 40 30 20

_____ °F

3.
9:00 P.M.

°F 100 90 80 70 60 50 40 30 20

_____ °F

4. How much warmer is it at 12:30 P.M. than at 8:00 A.M.?

_____ °F warmer

5. How much colder is it at 9:00 P.M. than at 5:00 P.M.?

_____ °F colder

Add. Remember to practice and memorize your addition facts.

6. 4 + 2 + 6 = _____

7. _____ = 7 + 2 + 7

8. _____ = 9 + 3 + 1

9. 8 + 4 + 5 = _____

10. 5 + 9 + 4 = _____

11. 7 + 8 + 3 = _____

Name Date Time

Practice Set 31

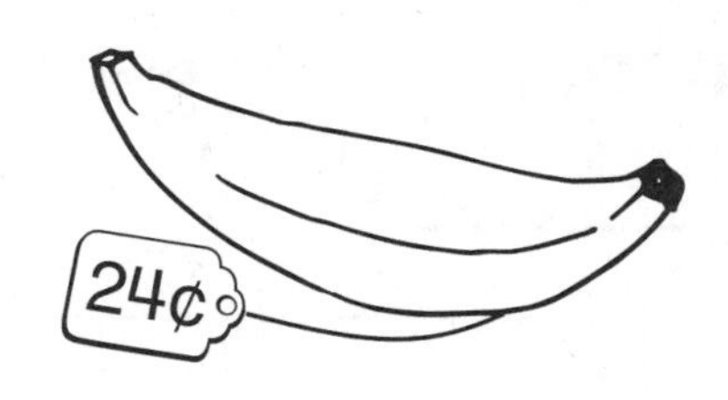

1. **Writing/Reasoning** You have $1.00. Which two different items can you buy? Explain your answer.

____

____

____

Fill in the diagram. Then write a number model.

2. Mia has 30 dimes. She gives 10 dimes to her brother. How many dimes does Mia have now?

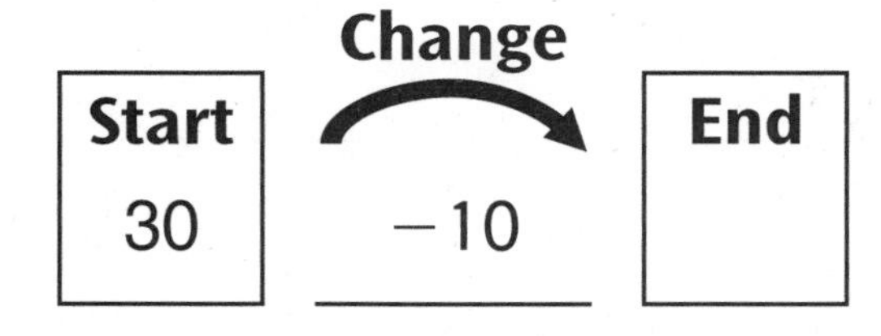

Number model: ______________

3. Mark bakes 12 peanut butter cookies and 24 oatmeal raisin cookies. How many cookies does he bake?

Total	
Part	**Part**
12	24

Number model: ______________

Name _______________ Date _______ Time _______

Practice Set 32

How long is each line segment?

1.

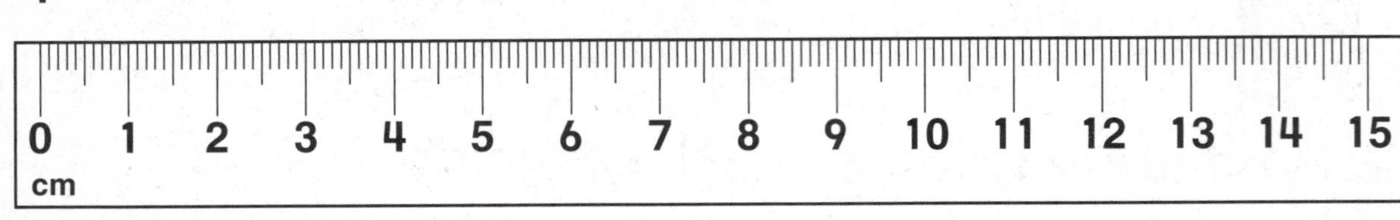

about _____ centimeters

2.

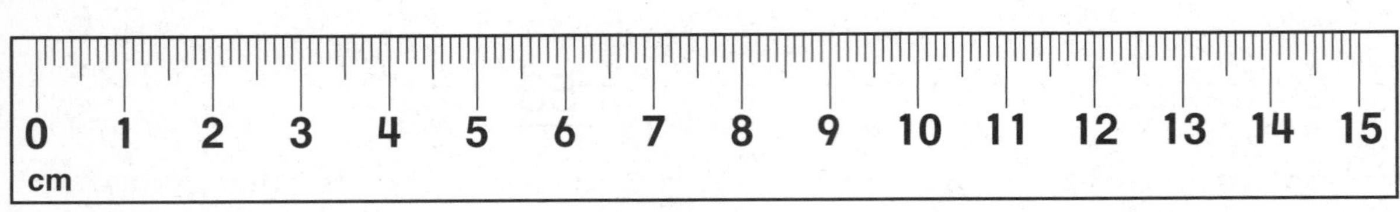

about _____ centimeters

Use these numbers to answer the questions below:

429 539 421 527 581

3. Which numbers have 9 ones? _______________

4. Which numbers have 5 hundreds? _______________

5. Which numbers have 2 tens? _______________

6. Which numbers have 4 hundreds? _______________

7. **Writing/Reasoning** How do you know that 539 is larger than 527? Explain your answer.

____

____

Name Date Time

Practice Set 33

Use the partial-sums method to solve the problems.

Example

10s	1s	
5	7	
+ 2	5	
7	0	Add the tens (50 + 20 = 70).
+ 1	2	Add the ones (7 + 5 = 12).
8	2	Combine the tens and the ones (70 + 12 = 82).

1. $\begin{array}{r} 56 \\ +\ 37 \\ \hline \end{array}$

2. $\begin{array}{r} 77 \\ +\ 17 \\ \hline \end{array}$

3. $\begin{array}{r} 39 \\ +\ 53 \\ \hline \end{array}$

4. $\begin{array}{r} 237 \\ +\ 154 \\ \hline \end{array}$

How long is each line segment?

5. = about ____ inches

0 1 2 3 4 5 6
inches

6. = about ____ inches

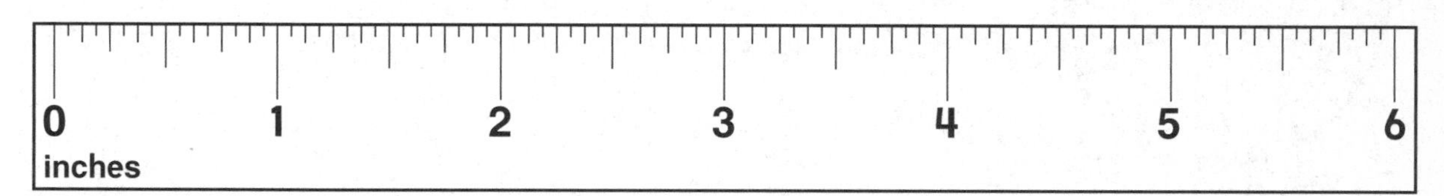

Name ____________ Date ________ Time ________

Practice Set 34

Match.

1.

11:20

2.

6:35

3.

8:10

4.

Write two 3-digit numbers that have ...

5. 3 in the ones place.

________, ________

6. 7 in the hundreds place.

________, ________

7. 6 in the tens place.

________, ________

8. 1 in the ones place and 0 in the tens place.

________, ________

9. Write six numbers using the digits 4, 9, and 2.

________, ________

________, ________

________, ________

Name Date Time

Practice Set 35

1. Find the figure on the right that has the same **shape** as the one below. Color it **red.**

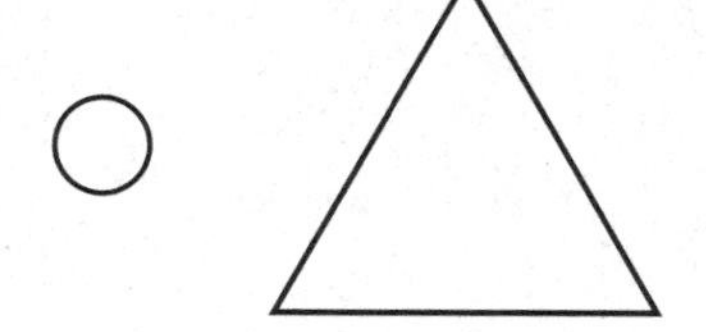

2. Find the figure on the right that is almost the same **size** as the one below. Color it **blue.**

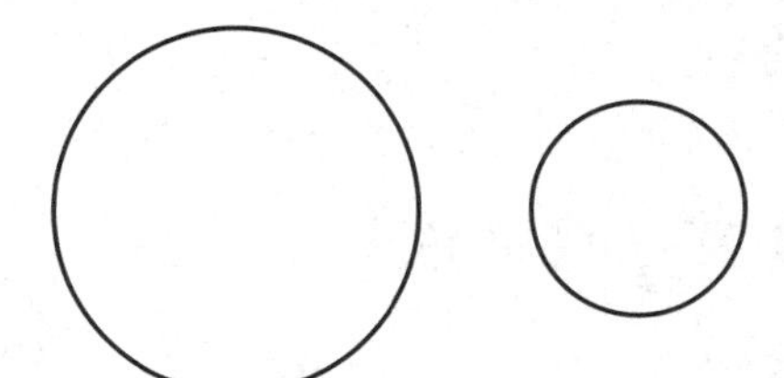

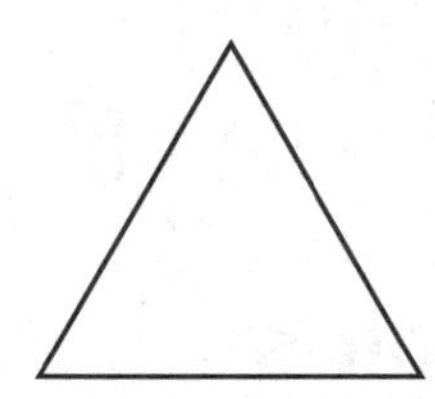

3. **Writing/Reasoning** Can a triangle have parallel line segments? Explain your answer.

____

____

Write <, >, or =.

4. 473 ☐ 437

5. 4 + 5 ☐ 9

6. 55 ☐ 48 + 5

7. 87¢ ☐ $0.86

8. 5 nickels ☐ 1 quarter

9. 4 + 66 ☐ 66 + 40

Name Date Time

Practice Set 36

How many?

1. 

How many rows? ______

How many in each row? ______

How many in all? ______

2. 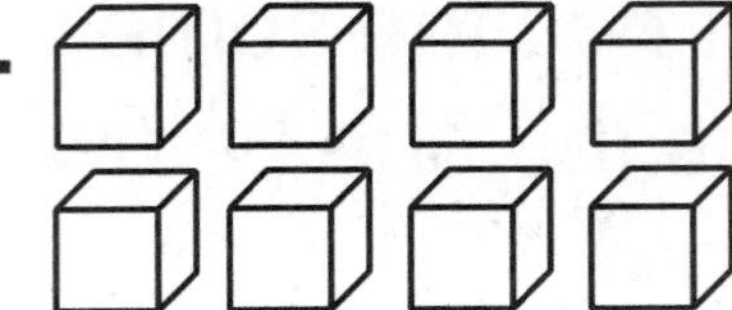

How many rows? ______

How many in each row? ______

How many in all? ______

Use Ⓠs, Ⓓs, Ⓝs, or Ⓟs to show the coins you could use to buy each item.

3. 

4. 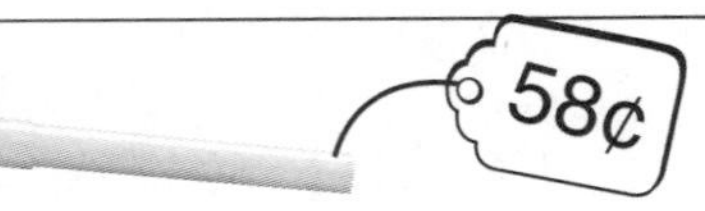

5. 

Name Date Time

Practice Set 37

Write the name of each quadrangle.

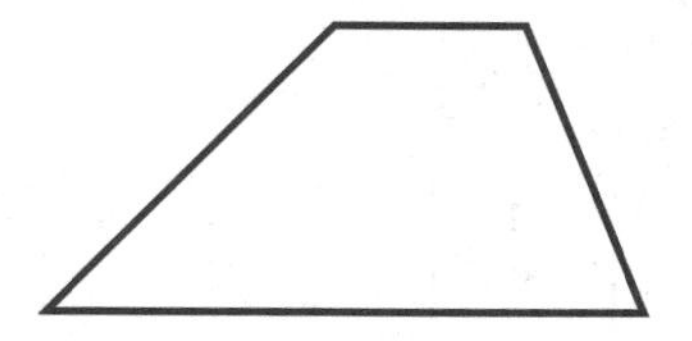

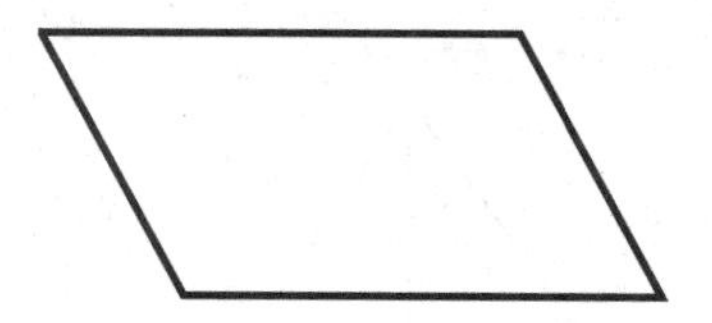

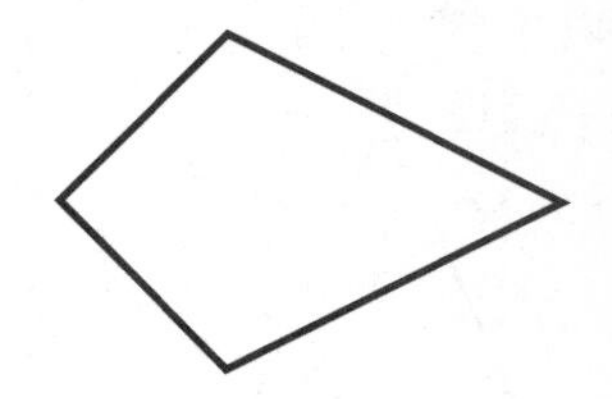

1. ______________ 2. ______________ 3. 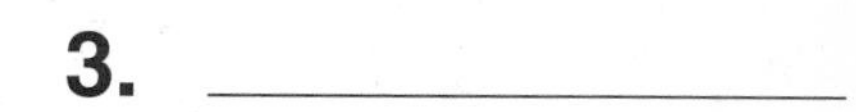

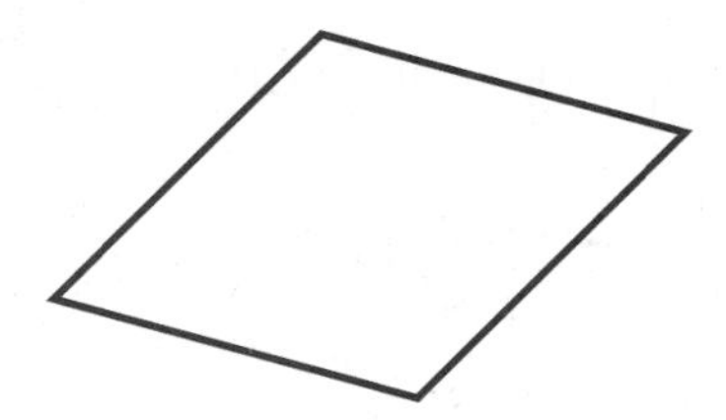

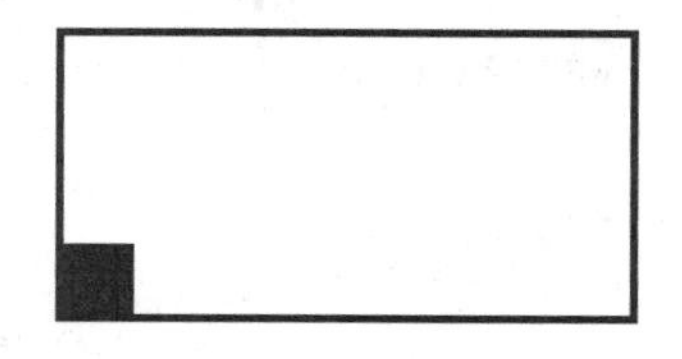

4. ______________ 5. ______________ 6. ______________

7. Draw a black box to show each right angle in the quadrangles above. One box has been drawn for you in Figure 5.

8. Color each pair of parallel lines **green.** If a shape has more than one pair of parallel lines, color the second pair **yellow.**

9. **Writing/Reasoning** Name something that you are certain will happen and tell how you know.

____

____

Name Date Time

Practice Set 38

How many edges, faces, and vertices?

Example

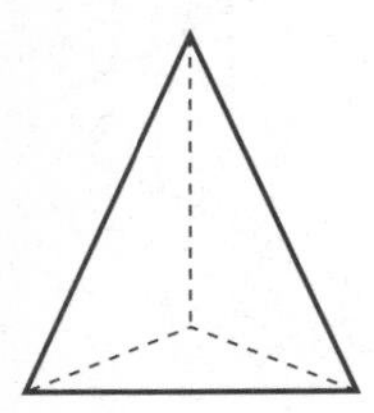

Triangular Pyramid

Edges: 6

Faces: 4

Vertices: 4

1.

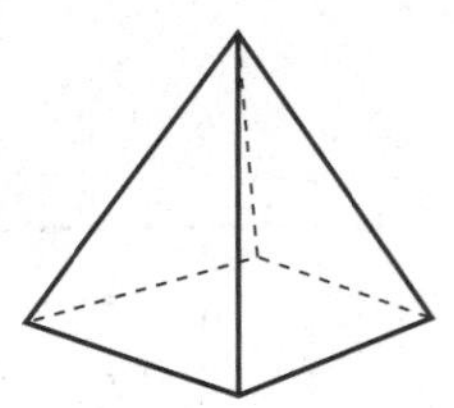

Rectangular Pyramid

Edges: ____

Faces: ____

Vertices: ____

2.

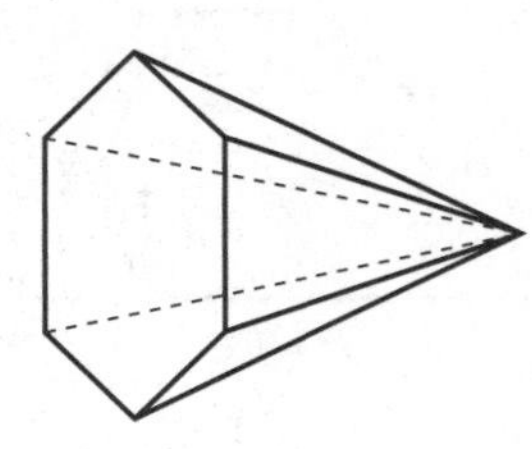

Hexagonal Pyramid

Edges: ____

Faces: ____

Vertices: ____

Write 8 names for each number.

3.

4.

Name Date Time

Practice Set 39

Draw all the lines of symmetry for each shape.
How many lines of symmetry does each shape have?

1. 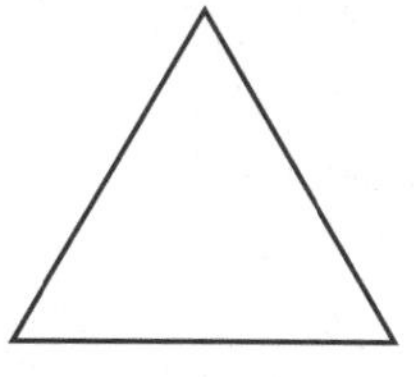

____ lines of symmetry

2. 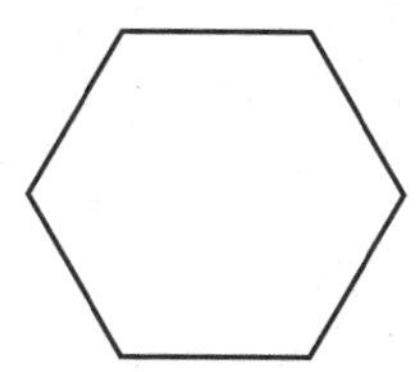

____ lines of symmetry

Draw the other half of each shape.
Then name the shape.

3. 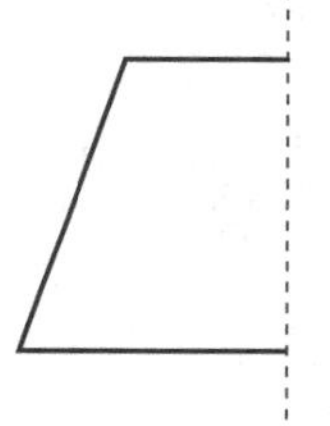

____

4. 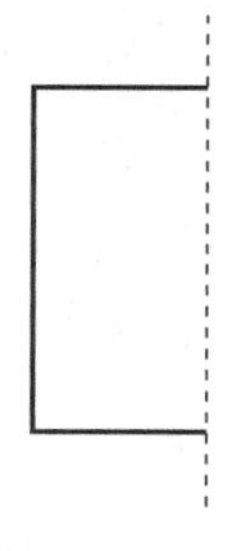

____

5. **Writing/Reasoning** How many lines of symmetry does a circle have? Explain your answer.

____

____

Add. Remember to practice and memorize your addition facts.

6. 124 + 133 = ______

7. 237 + 102 = ______

8. 100 + 73 = ______

9. 99 + 50 = ______

Name Date Time

Practice Set 40

Solve.

1. 15 + 5 + 2 = ______
2. 13 + 2 + 7 = ______
3. 6 + 50 + 25 = ______
4. 18 + 4 + 9 = ______
5. 21 + 9 + 1 = ______
6. 12 + 7 + 8 = ______
7. 17 + 8 + 3 = ______
8. 24 + 6 + 2 = ______
9. 28 + 5 + 2 = ______
10. 19 + 9 + 1 = ______

Write the fact family.

11.

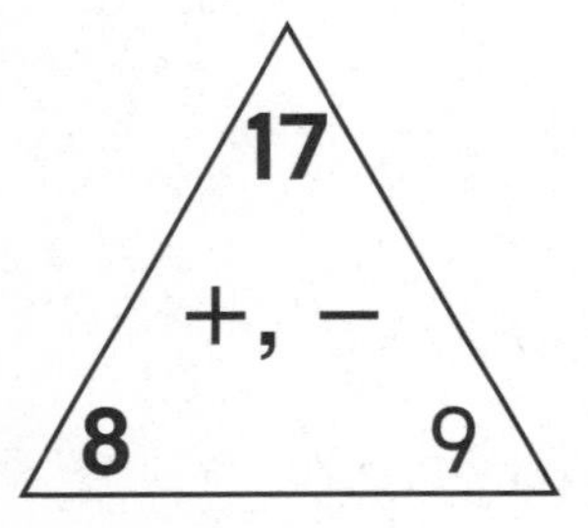

______
______
______
______

12.

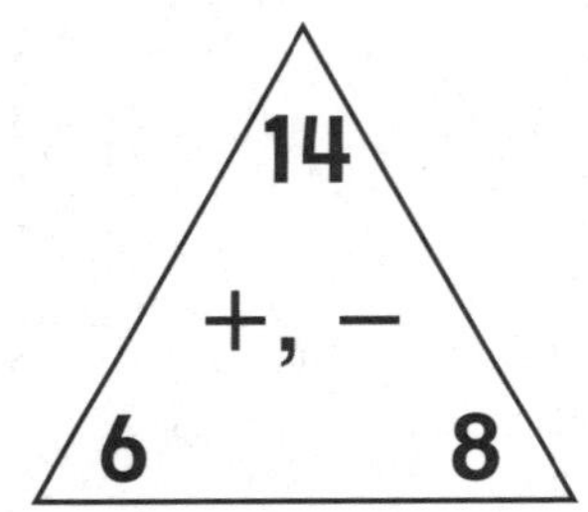

______
______
______
______

13.

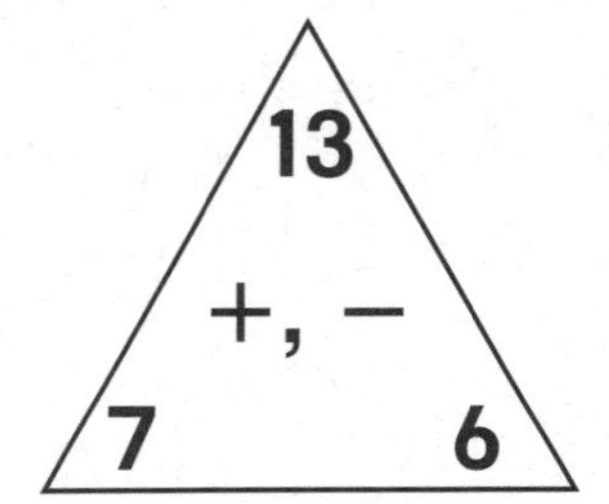

______
______
______
______

Write two facts for each fact family.

14. 9, 9, 18 ____ + ____ = ____ ____ − ____ = ____

15. 10, 5, 5 ____ + ____ = ____ ____ − ____ = ____

16. 8, 8, 16 ____ + ____ = ____ ____ − ____ = ____

17. 14, 7, 7 ____ + ____ = ____ ____ − ____ = ____

Use with or after Lesson 6•1.

Name Date Time

Practice Set 41

Finish each diagram. Then write a number model.

Example

Quantity
59

Quantity
32

27 **Difference**

59 − 32 = 27

1.

Quantity
86

Quantity

59 **Difference**

______

2.

Quantity
92

Quantity
34

______ **Difference**

______

3.

Quantity

Quantity
14

56 **Difference**

______

Find the difference between the temperatures.

4. 34°C and 15°C ______

5. 13°F and 67°F ______

6. 95°F and 57°F ______

7. 24°C and 9°C ______

Subtract. Remember to practice and memorize your subtraction facts.

8. 18 − 8 = ______ **9.** 7 − 4 = ______ **10.** 13 − 9 = ______

Name Date Time

Practice Set 42

1. Finish the bar graph.
2. How many more dogs than cats are owned by second graders?

   ______ more dogs
3. How many more fish than birds are owned as pets?

   ______ more fish

Pets Owned by Second Graders

Cat	𝍸 𝍸 𝍸 𝍸 𝍸 𝍷𝍷𝍷𝍷
Fish	𝍸 𝍸 𝍸 𝍷𝍷
Dog	𝍸 𝍸 𝍸 𝍸 𝍸 𝍸 𝍸 𝍷𝍷𝍷
Bird	𝍸 𝍷𝍷𝍷

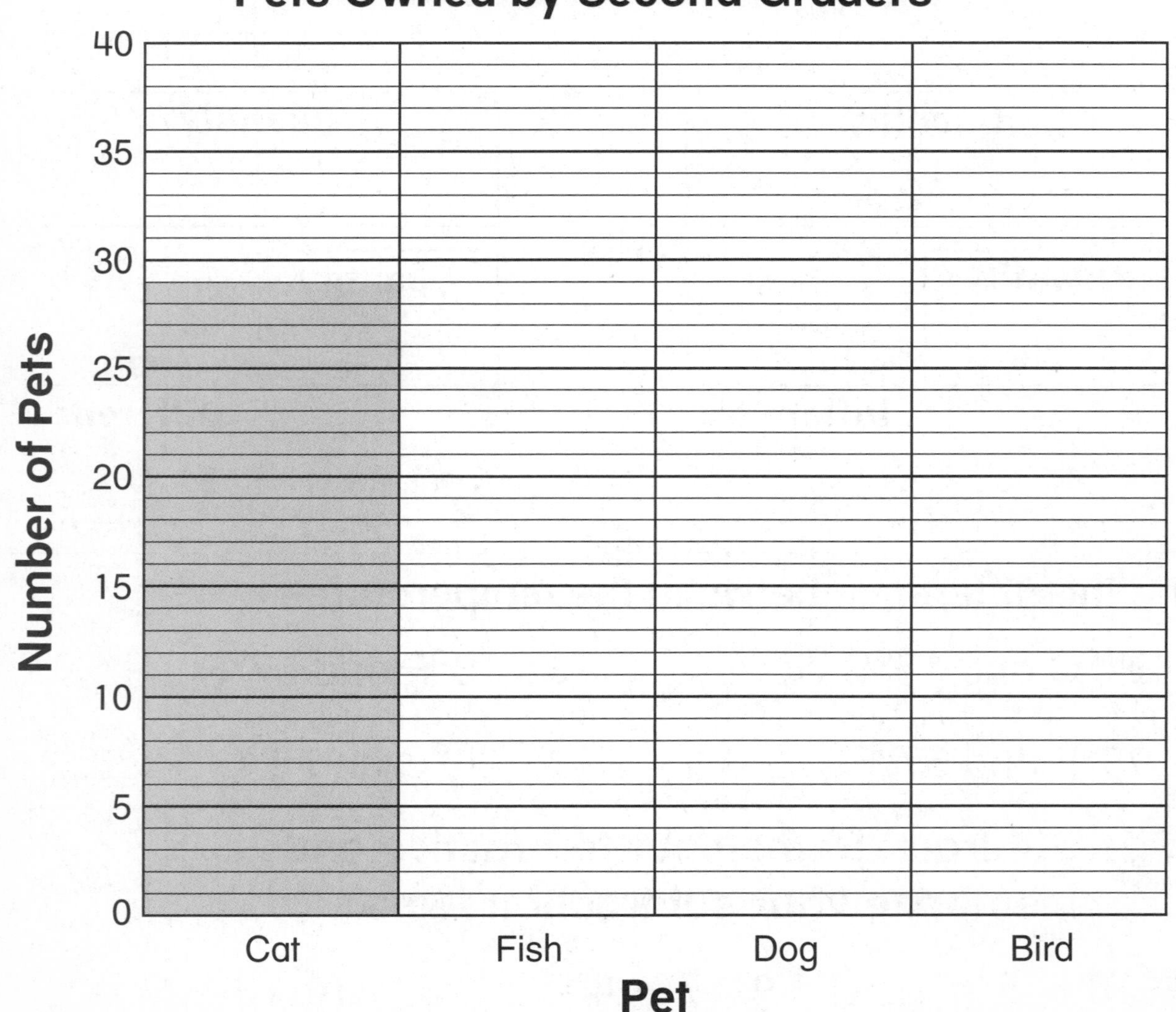

Name Date Time

Practice Set 43

Use a diagram to help you solve each problem. Write a number model to go with each problem.

1. One duck has 13 ducklings. Another duck has 12. How many ducklings are there in all?

   There are ____ ducklings in all.

   Number model:

   ____

Quantity

Quantity

Difference

2. 65 worker bees were in the hive. 23 of them flew away. How many worker bees are left in the hive?

   ____ worker bees are left.

   Number model:

   ____

Start | Change | End

3. 39 roses were blooming in the garden. Alberto picked 12 of them. How many roses were not picked?

   ____ roses were not picked.

   Number model:

   ____

Total	
Part	Part

Name Date Time

Practice Set 44

Make a ballpark estimate. Then subtract. Trade first if you need to.

1.

longs 10s	cubes 1s
5	11
~~6~~	~~1~~
− 3	2

Answer

Ballpark estimate:

60 − 30 = 30

2.

longs 10s	cubes 1s
4	5
−1	7

Answer

Ballpark estimate:

____

Find the rule. Complete the frames.

3.

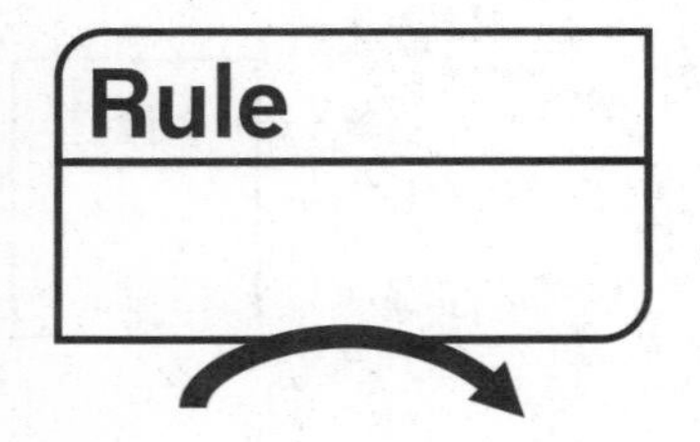

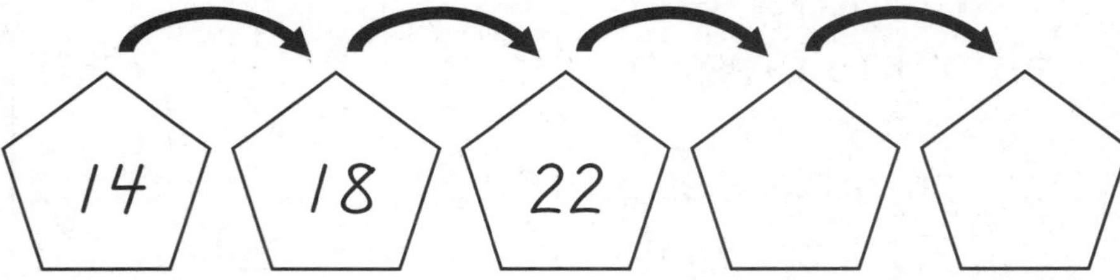

4.

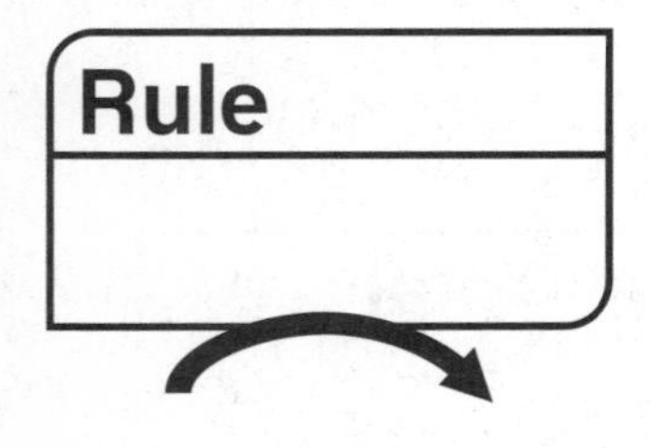

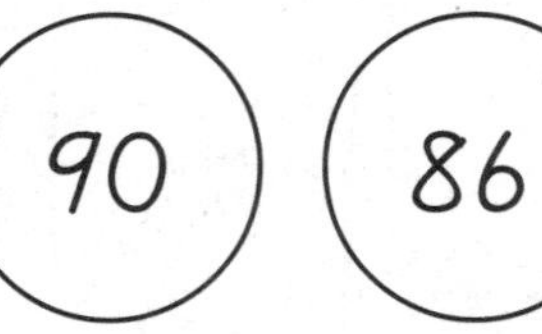

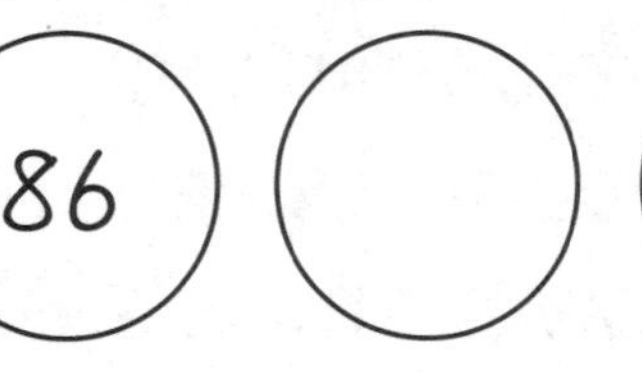

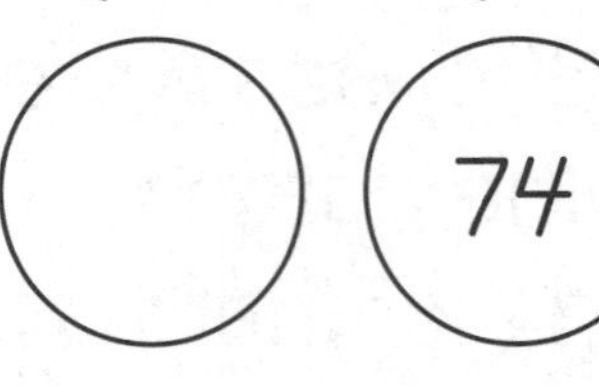

5.

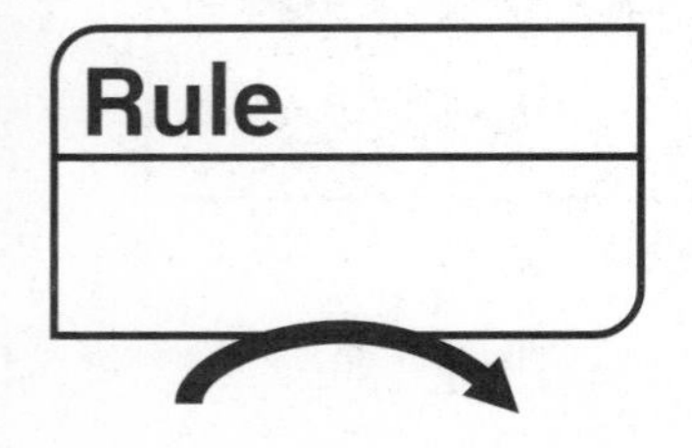

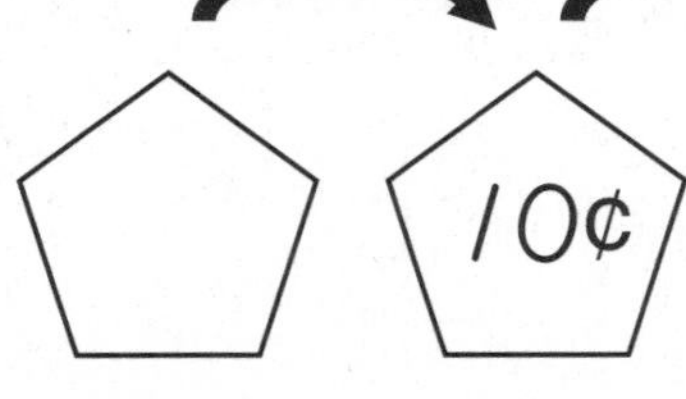

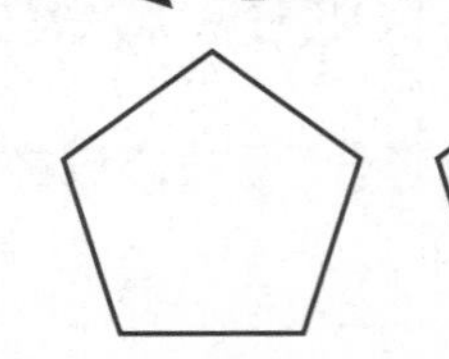

Name Date Time

Practice Set 45

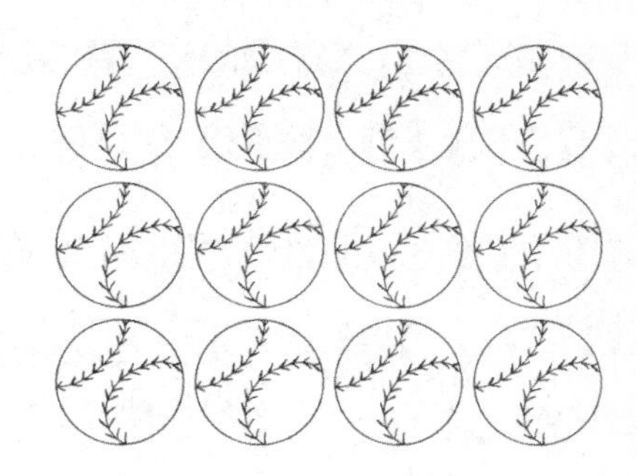

1. How many rows? ______

   How many in each row? ______

   How many balls in all? ______

2. How many nuts in all? ______

   How many bowls? ______

   How many nuts in each bowl if each bowl gets the same amount? ______

Write the rule. Then fill in the table.

3.

in	out
20	30
70	80
	90
60	

4.

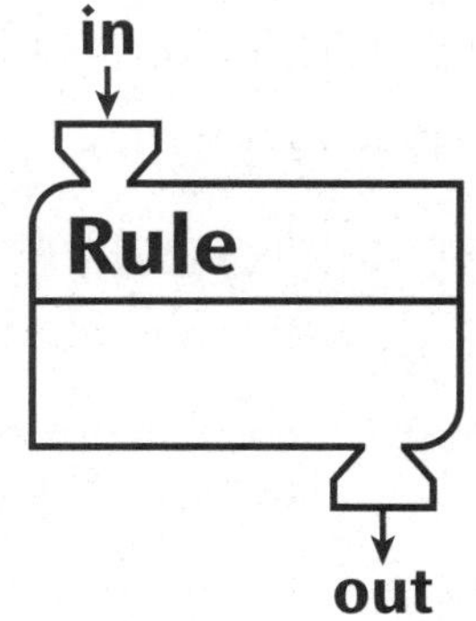

in	out
24	16
38	30
16	
	20

5.

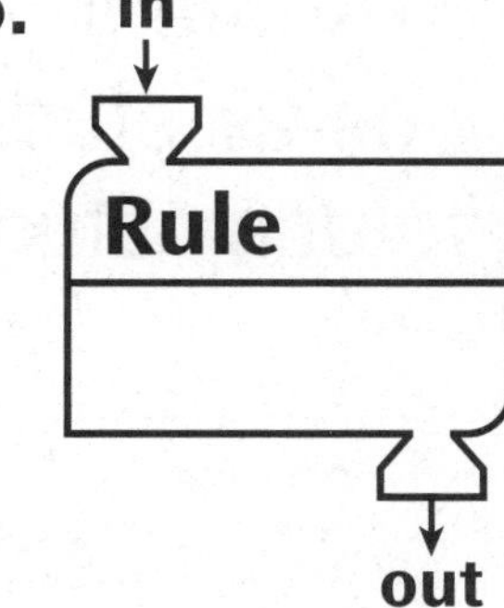

in	out
12	18
14	20
17	
	12

Name Date Time

Practice Set 46

Solve each problem. Draw pictures or use counters to help you.

1. Jim has 3 boxes of colored pencils. There are 8 colored pencils in each box. How many colored pencils does he have in all?

Answer: ______ colored pencils

2. Ms. Hughes brought 5 bags of oranges to school. Each bag contained 9 oranges. How many oranges did Ms. Hughes bring in all?

Answer: ______ oranges

3. **Writing/Reasoning** Ms. Hughes wants to share the oranges with her class. She has 21 students. Can she give 2 oranges to each student? Explain your answer.

____

____

____

Write *even* or *odd*.

4. 203 ______ **5.** 2,221 ______ **6.** 198 ______

7. 64 ______ **8.** 32 ______ **9.** 287 ______

Use with or after Lesson 6•7.

Name Date Time

Practice Set 47

Draw pictures or use counters to solve each problem.

1. Rosemary put 4 vases of flowers on the tables. Each vase has 5 flowers. How many flowers are there in all?

   Answer: _____ flowers

2. On Saturday, 6 children went fishing. Each child caught 4 fish. How many fish did the children catch in all?

   Answer: _____ fish

3. Dave has 2 boxes of crayons. Each box has 8 crayons. How many crayons does Dave have in all?

   Answer: _____ crayons

4. Karen bought 3 packs of balloons. Each pack had 5 balloons. How many balloons did Karen buy in all?

   Answer: _____ balloons

5. **Writing/Reasoning** Write your own multiplication story. Then find the answer.

   ____

   Answer: __________

Name Date Time

Practice Set 48

For each problem
- **make an array**
- **fill in the number model**
- **write the answer**

Example The garden has 5 rows of flowers. Each row has 3 flowers. How many flowers are there?

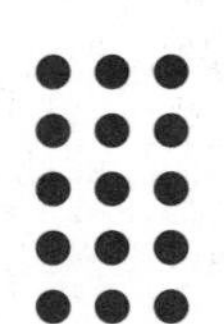

5 × 3 = 15

There are 15 flowers.

1. The gameboard has 3 rows. Each row has 6 squares. How many squares are there?

____ × ____ = ____

There are ____ squares.

2. The calculator has 4 rows of keys. There are 5 keys in each row. How many keys are there?

____ × ____ = ____

There are ____ keys.

3. Each page of an album has 2 rows of pictures. Each row has 3 pictures. How many pictures are on each page?

____ × ____ = ____

There are ____ pictures.

Add.

4. 10 = 3 + ______

5. ______ + 28 = 40

6. 45 + ______ = 70

7. 90 = 67 + ______

Name Date Time

Practice Set 49

Solve.

1. 20 cents shared equally ...

by 4 people	**by 5 people**	**by 6 people**
____¢ per person	____¢ per person	____¢ per person
____¢ remaining	____¢ remaining	____¢ remaining

2. 35 books shared equally ...

by 5 children	**by 6 children**
____ books per child	____ books per child
____ books remaining	____ books remaining

Count by 2s.

3. 900, 902, ____, ____, ____, ____, ____, ____, ____, ____

4. 389, ____, ____, ____, ____, ____, 401, ____, ____, ____

Count by 10s.

5. 610, 620, ____, ____, ____, ____, ____, ____, ____, ____

6. 716, 706, ____, ____, ____, ____, ____, ____, ____, ____

7. Circle each pair of parallel lines.

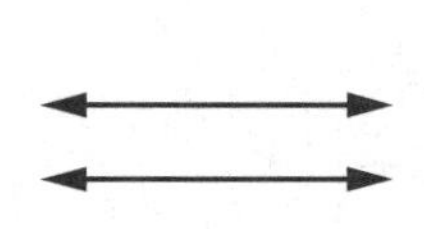
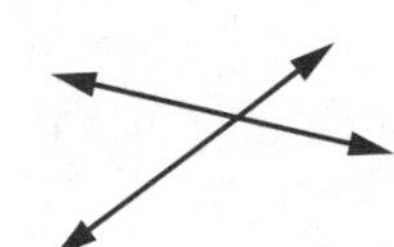
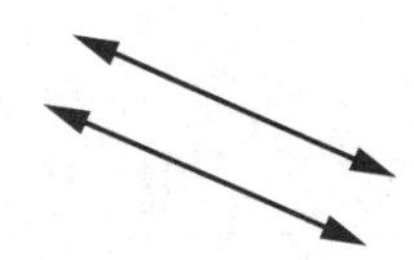
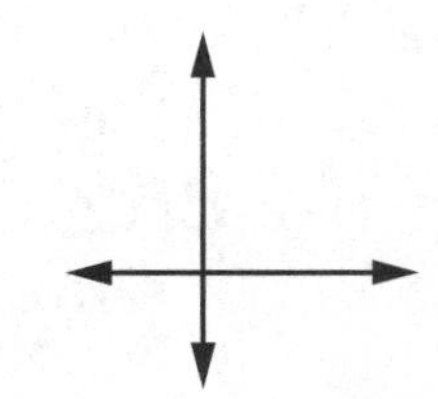

Name Date Time

Practice Set 50

801	802	803							810
811			814					819	
	822				826	827			830
		833		835				839	840
			844				848		
851					856				
	862							869	
871									880
		883	884			887			
891	892				896		898		900
	902	903						909	
	912				916				920

1. Fill in all of the missing numbers on the grid.
2. Now start at 804 and count by 10s.
   Put an **X** over each number as you count.
3. Now circle three *odd* numbers on the grid.

Use the grid to find these sums.

4. 803 + 10 = ______
5. ______ = 30 + 844
6. ______ = 862 + 40
7. 40 + 869 = ______

Name Date Time

Practice Set 51

Fill in the missing number.

1. 50 = 42 + ____

2. 90 = ____ + 89

3. 64 + ____ = 70

4. 30 = 27 + ____

5. 35 + ____ = 40

6. 53 + ____ = 60

For Problems 7–12, draw the shape (⏢, ○, □, or △) for the number.

601	602	603		△	□	607	608	609	610
611				615			○		620
621	622	□			626			△	
	632	633		⏢		637			640
	642		△				648	649	△
651	○			655			○		660
		663			□			669	○
671			674			677	678		
⏢		683			686			689	△
	692			□				699	△
	702				706				
711		713				○			720

7. 618 ________

8. 695 ________

9. 666 ________

10. 681 ________

11. 700 ________

12. 717 ________

Name Date Time

Practice Set 52

Add.

1.	2.	3.	4.
20	17	5	16
12	9	13	14
+ 6	+ 18	+ 8	+ 4

Use your calculator to find the answer.

Example

Enter 24.

Change to 50.

Add or subtract? add

How much? 26

5. Enter 90.

Change to 67.

Add or subtract? ______

How much? ______

6. Enter 70.

Change to 22.

Add or subtract? ______

How much? ______

7. Enter 48.

Change to 80.

Add or subtract? ______

How much? ______

8. **Writing/Reasoning** In Problem 7, how did you know whether to add or subtract?

____

____

Name Date Time

Practice Set 53

Fill in each table.

1. Rule: Double

in	out
3	6
4	
5	
6	

2. Rule: Halve

in	out
48	24
60	
18	
12	

Solve.

Animal Measurements

Eagle
length 26 in.

Beaver
length 30 in.

Raccoon
weight 25 lb

Koala
weight 20 lb

3. How much longer is the beaver than the eagle?

______ in. longer

4. How much do the koala and the raccoon weigh together?

______ lb

Name Date Time

Practice Set 54

Read the scale. Tell the weight.

1.

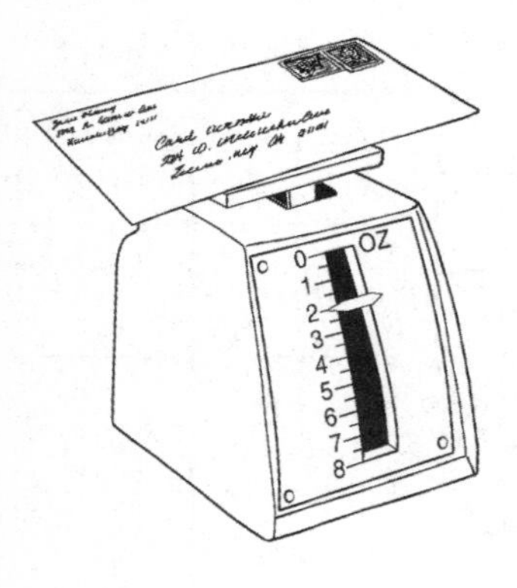

2.

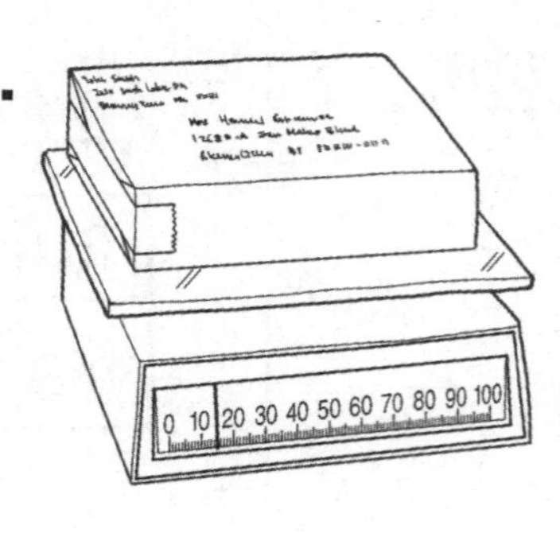

3.

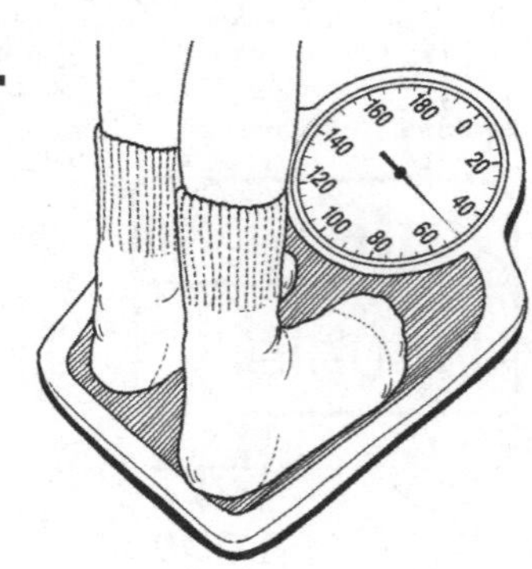

1. ____ ounces (oz)
2. ____ pounds (lb)
3. ____ pounds (lb)

Draw pictures or use counters to find the answers.

4. 15 toys shared equally by 4 friends

_____ toys per friend

_____ toys remaining

5. 29 bones shared equally by 9 dogs

_____ bones per dog

_____ bones remaining

Find the missing number.

6. 59 = _____ + 9

7. _____ + 40 = 63

8. _____ + 60 = 72

9. 93 = 50 + _____

10. 70 + _____ = 99

11. 55 = _____ + 20

Name Date Time

Practice Set 55

1. Measure the eraser to the nearest $\frac{1}{2}$ inch.

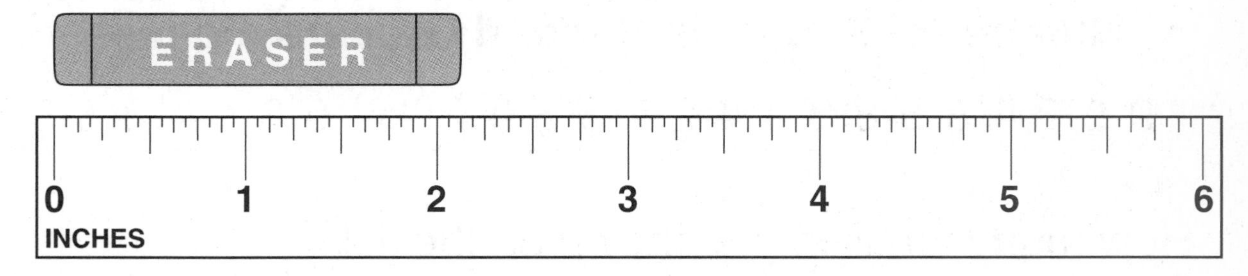

The eraser is about _______ inches long.

2. Measure the crayon to the nearest centimeter.

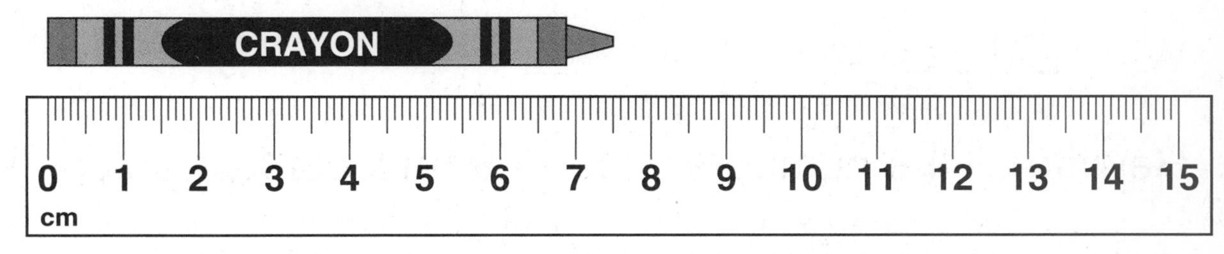

The crayon is about _______ centimeters long.

3. **Writing/Reasoning** Two of your friends measure your arm span. One uses an inch ruler. The other uses a centimeter ruler. Which one will get a greater number? Explain your answer.

____

____

____

Add or subtract.

4. $\begin{array}{r} 33 \\ +\ 19 \\ \hline \end{array}$

5. $\begin{array}{r} 87 \\ -\ 56 \\ \hline \end{array}$

6. $\begin{array}{r} 65 \\ -\ 38 \\ \hline \end{array}$

7. $\begin{array}{r} 43 \\ +\ 59 \\ \hline \end{array}$

Name Date Time

Practice Set 56

Use this list of numbers to answer each question below:

Number of People in Second Graders' Families
2 5 3 4 6 5 3 3 4 4 2 7 3 5 4 2 4 4 5 3 6 4 3 6 5

1. How many families are shown on the list? ___________

2. What is the greatest number of people in a second grader's family? ___________

3. What is the least? ___________

4. Rewrite all the numbers in order from least to greatest.

____

____

5. How many families have 3 people? ___________

6. What is the middle, or **median,** value shown on the list? ___________

Solve.

7. A pencil costs 8¢. A crayon costs 4¢. How much do they cost together? ________

8. Gum costs $0.10. Candy costs $0.06. How much more does the gum cost? ________ more

Use with or after Lesson 7•7.

Name Date Time

Practice Set 57

Use the frequency table to answer each question.

1. How many second graders' heights are shown in the table?

____

2. What is the middle, or **median,** value shown in the table?

____

Heights of Second Graders

Height (in.)	Frequency	
	Tallies	Number
42	///	3
43	~~////~~	5
44	~~////~~ /	6
45	~~////~~ ///	8
46	//	2

Solve.

3. How many dots are in this array? Count by 3s.

● ● ● ● ● ● ● ● ●
● ● ● ● ● ● ● ● ●
● ● ● ● ● ● ● ● ●

_________ dots

4. Describe this array:

● ● ● ● ● ●
● ● ● ● ● ●

_________ by _________

Draw the other half of each figure.

5.

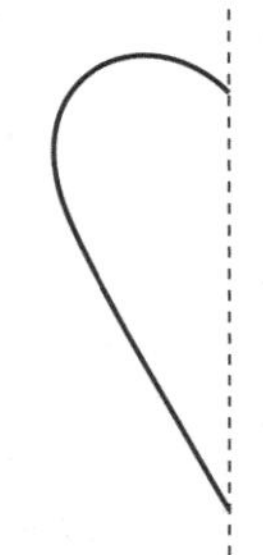

6.

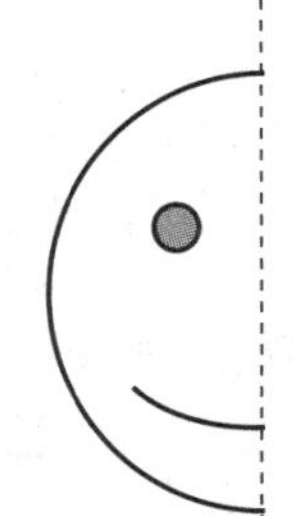

7.

Name Date Time

Practice Set 58

Write 2 fractions for each shape.

Example

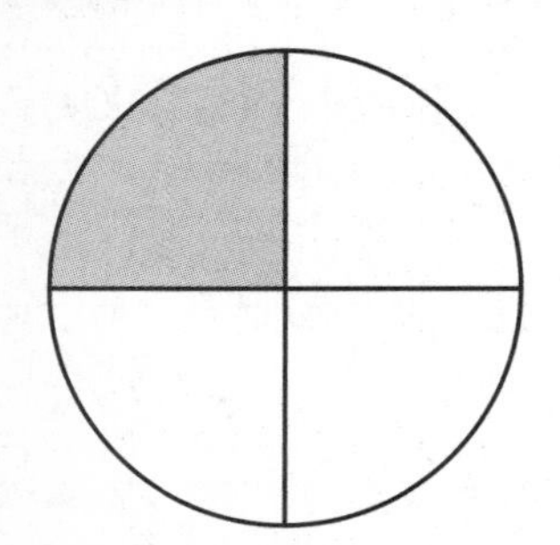

4 equal parts

part shaded = $\frac{1}{4}$

part unshaded = $\frac{3}{4}$

1.

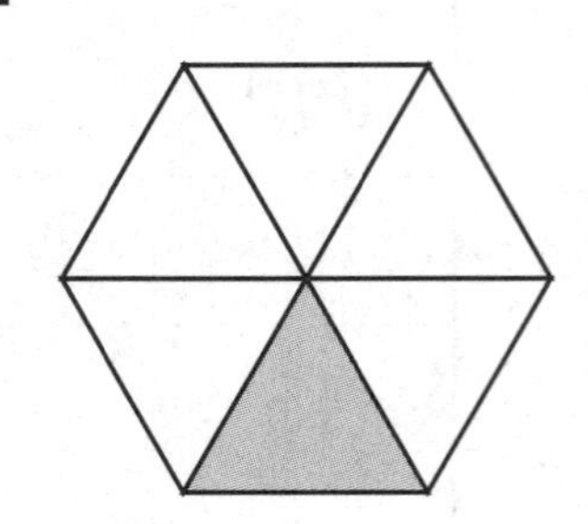

6 equal parts

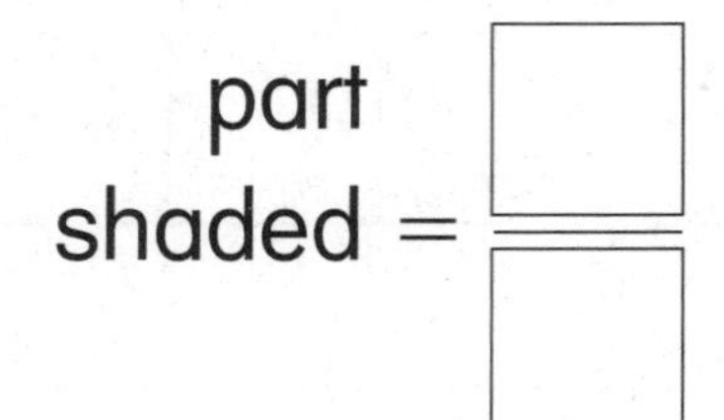

part shaded = ____ / ____

part unshaded = ____ / ____

2.

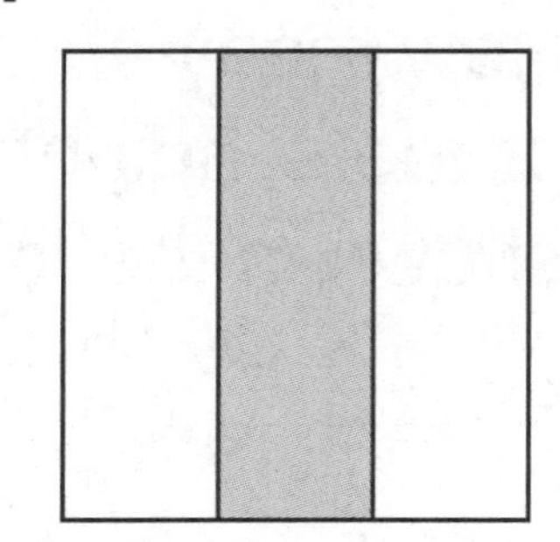

3 equal parts

part shaded = ____ / ____

part unshaded = ____ / ____

Write the missing numbers.

3. 80 + ____ = 92

70 + ____ = 92

60 + ____ = 92

50 + ____ = 92

40 + ____ = 92

30 + ____ = 92

4. 87 = ____ + 10

77 = ____ + 10

67 = ____ + 10

57 = ____ + 10

47 = ____ + 10

37 = ____ + 10

Use with or after Lesson 8•1.

Name Date Time

Practice Set 59

Label each part of each shape.

Example

$\frac{1}{6}$	$\frac{1}{6}$	$\frac{1}{6}$
$\frac{1}{6}$	$\frac{1}{6}$	$\frac{1}{6}$

1.

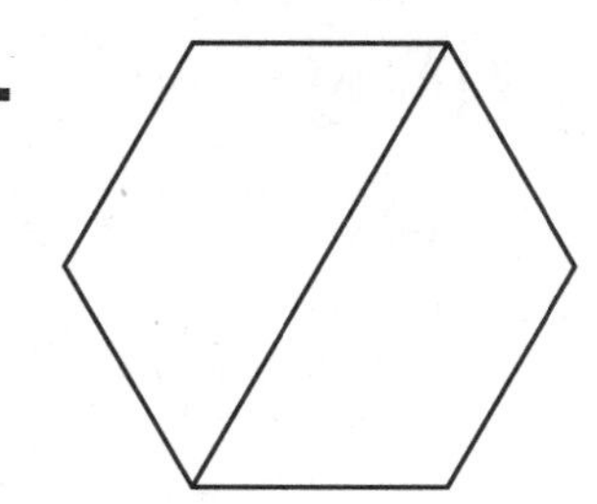

2.

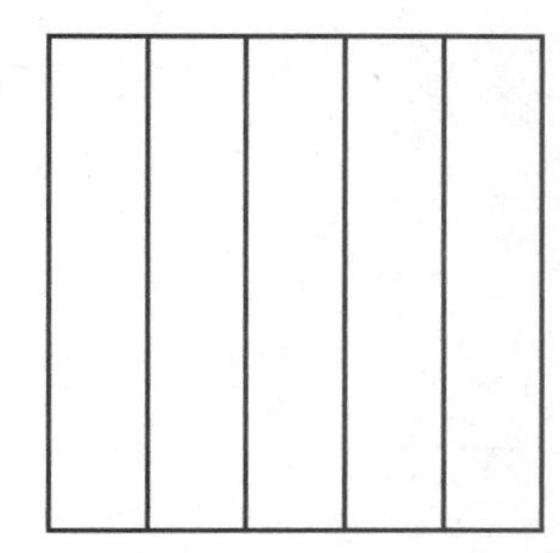

3.

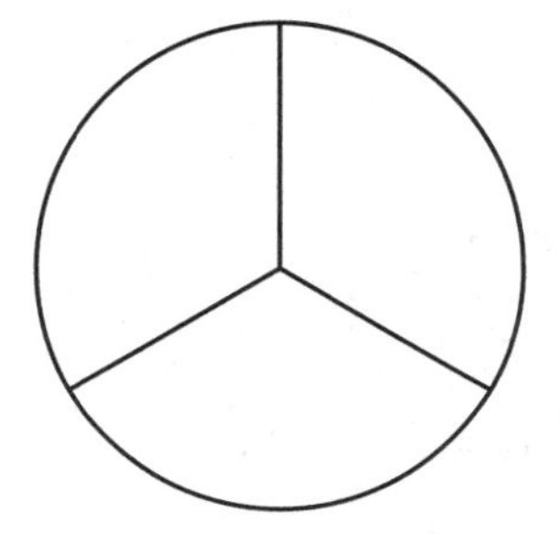

4.

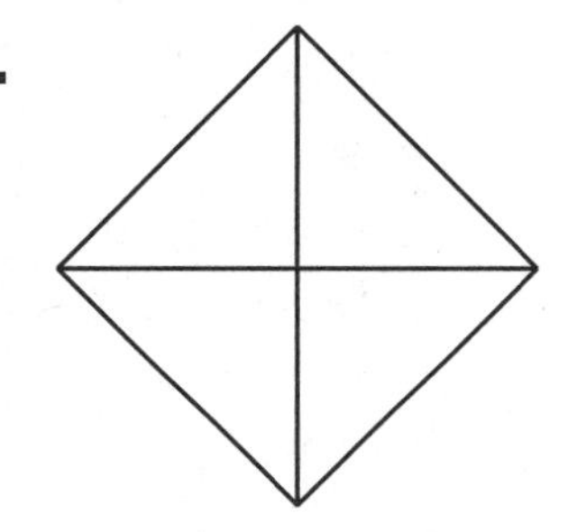

5.

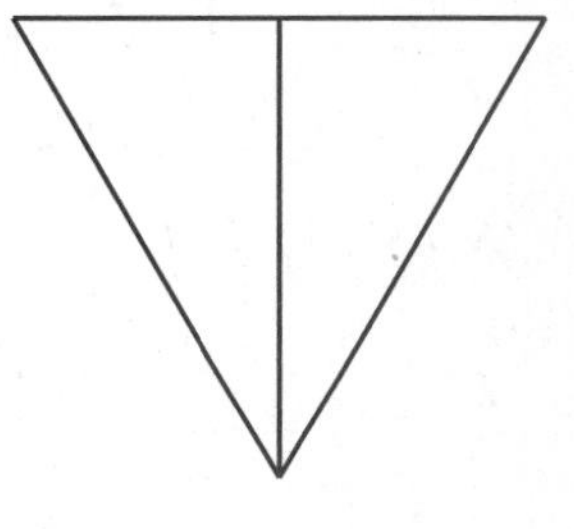

Count by 3s.

6. 82, 85, ____, ____, ____, ____, ____, ____, ____, ____

7. 216, ____, ____, 225, ____, ____, ____, ____, ____, ____

8. 803, ____, ____, ____, ____, ____, ____, ____, ____, 830

Add.

9. $29 + 4$

10. $65 + 9$

11. $37 + 6$

12. $52 + 7$

13. $49 + 8$

Name Date Time

Practice Set 60

Write the fraction for the shaded part.

Example

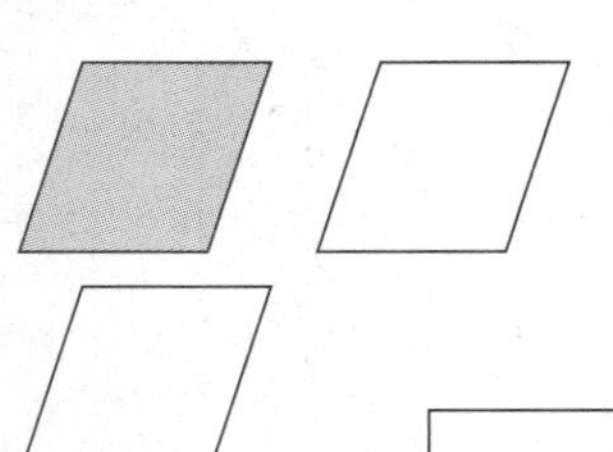

$\frac{1}{4}$

1. 

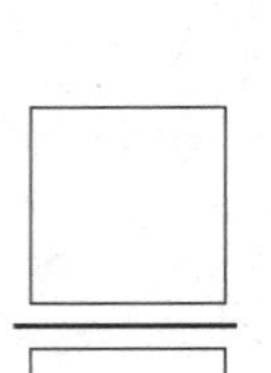

2. 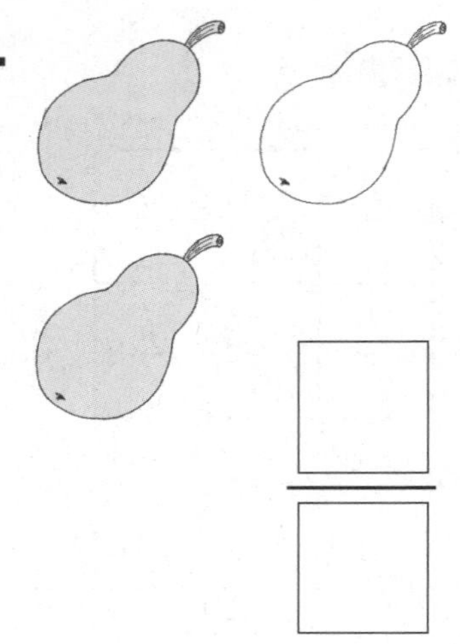

3. 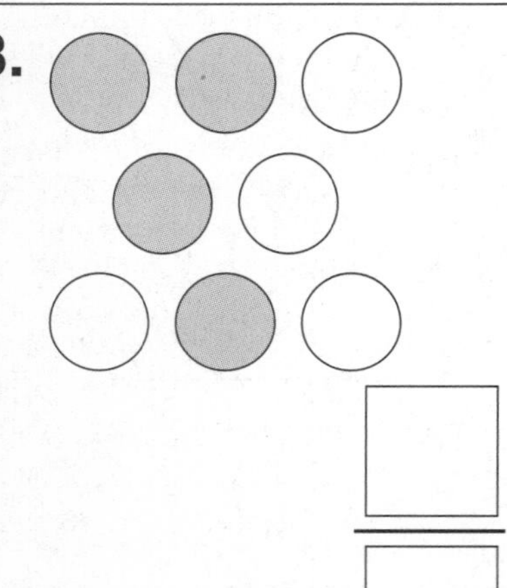

4. 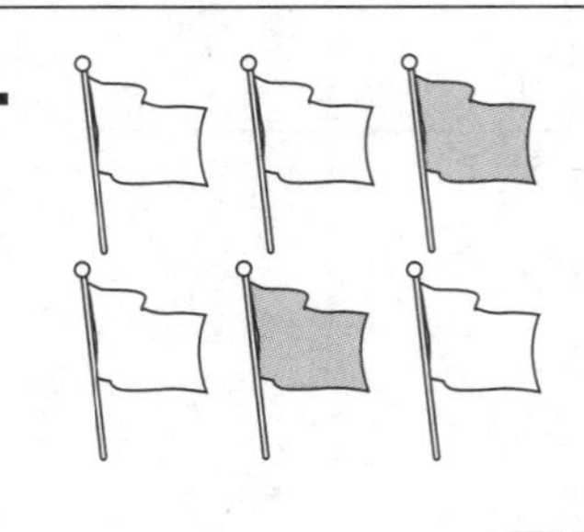

5. 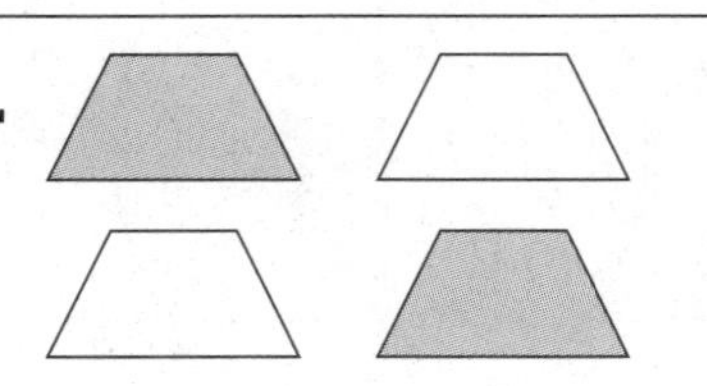

Count back by 10s.

6. 670, 660, ___, ___, ___, ___, ___, ___, 590, ___

7. 286, 276, ___, ___, ___, ___, ___, ___, ___, 196

8. 524, ___, ___, 494, ___, ___, ___, ___, ___, ___

9. 937, ___, ___, ___, ___, ___, 877, ___, ___, ___

Name Date Time

Practice Set 61

1. Shade $\frac{1}{4}$ of the rectangle.

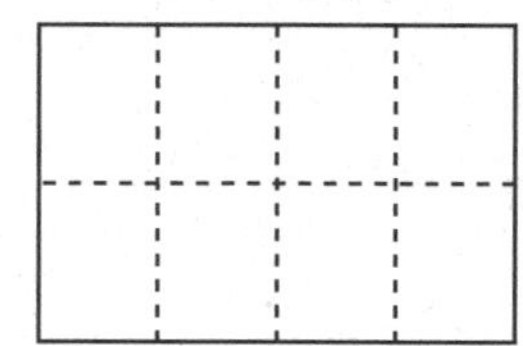

$\frac{1}{4} = \frac{\square}{8}$

2. Shade $\frac{1}{5}$ of the rectangle.

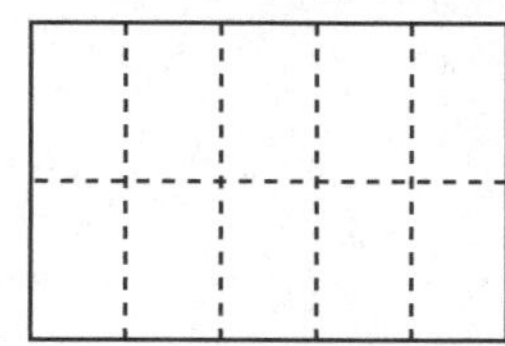

$\frac{1}{5} = \frac{\square}{10}$

3. Shade $\frac{1}{3}$ of the rectangle.

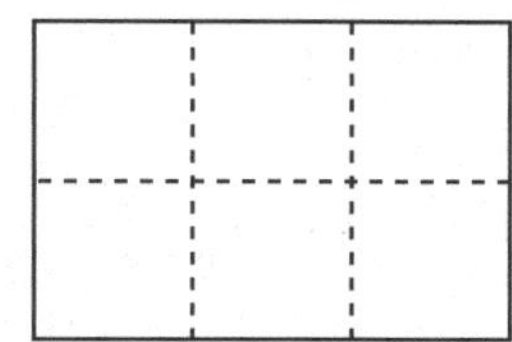

$\frac{1}{3} = \frac{\square}{6}$

4. Shade $\frac{1}{4}$ of the rectangle.

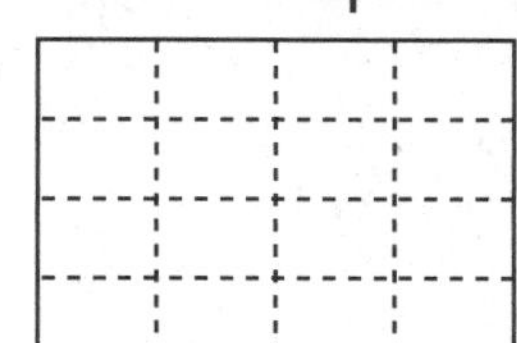

$\frac{1}{4} = \frac{\square}{16}$

5. **Writing/Reasoning** Compare the shaded parts in Problems 1 and 4. What do you notice?

____

____

Draw pictures or use counters to help you solve the problem.

6. 4 people share 16 pennies.

How many pennies does each person get? ____ pennies

$\frac{1}{4}$ of 16 pennies = ____ pennies.

$\frac{3}{4}$ of 16 pennies = ____ pennies.

Subtract.

7. 43 – 9 = _____ **8.** 54 – 8 = _____ **9.** 64 – 8 = _____

Name ______________________ Date ____________ Time ____________

Practice Set 62

Write <, >, or =.

1. This is ONE:

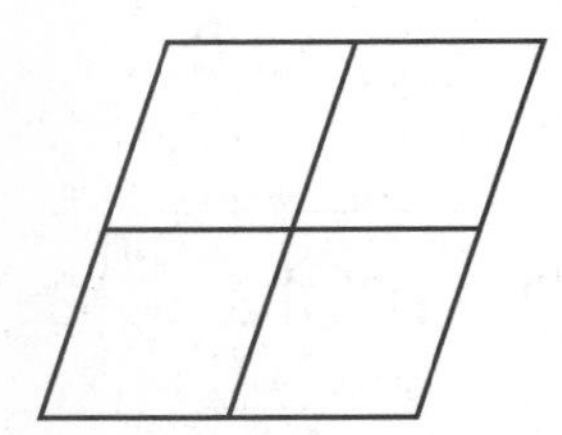

$\frac{2}{4} \square \frac{3}{4}$ $\frac{1}{2} \square \frac{2}{4}$ $\frac{3}{4} \square \frac{1}{4}$

2. This is ONE:

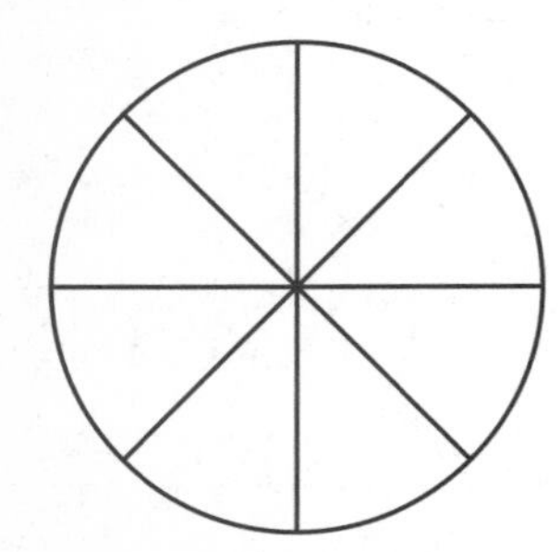

$\frac{5}{8} \square \frac{3}{4}$ $\frac{2}{4} \square \frac{3}{8}$ $\frac{1}{2} \square \frac{4}{8}$

Circle $\frac{1}{2}$ of each group. Write another name for $\frac{1}{2}$.

3. ● ● ● ● ● ● ● ● ● ● ● ● ________

4. ● ● ● ● ● ● ● ● ● ● ________

5. ● ● ● ● ________

6. ● ● ● ● ● ● ● ● ● ● ● ● ● ● ● ● ● ● ________

Write the missing number.

7. ______ + 5 = 10

8. 10 = ______ + 2

9. 80 = 53 + ______

10. 14 + ______ = 30

11. ______ + 59 = 90

12. 60 = ______ + 47

Name Date Time

Practice Set 63

Solve each problem.

1. 5 apples were on a tree. 2 apples fell to the ground. What fraction of the apples stayed on the tree?

   ______

2. Eight roses grew in a garden. $\frac{1}{4}$ of them were picked and put in a vase. How many roses are in the vase?

   ______

3. **Writing/Reasoning** Write your own fraction story for the picture below.

______

______

______

COMPUTATION PRACTICE **Add or subtract. Show your work in the space below.**

4. $\begin{array}{r} 84 \\ +\ 15 \\ \hline \end{array}$ **Answer** ______	5. $\begin{array}{r} 92 \\ -\ 51 \\ \hline \end{array}$ **Answer** ______	6. $\begin{array}{r} 74 \\ -\ 36 \\ \hline \end{array}$ **Answer** ______

Name Date Time

Practice Set 64

Count the unit squares to find the area of the rectangles.

1\.

_____ square units

2.

_____ square units

3.

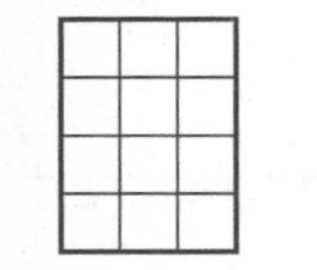

_____ square units

4.

_____ square units

Solve each problem. You can use counters or draw pictures.

5. There are 4 shelves on the wall. Each shelf holds 7 books. How many books are there in all?

Answer: _____ books

6. Pat bought 5 packs of juice boxes. Each pack holds 6 boxes. How many juice boxes did he buy?

Answer: _____ juice boxes

Name Date Time

Practice Set 65

How much is shaded? Write a fraction.

1.

0 1 2 3 4 5 6
INCHES

_______ in.

2.

0 1 2 3 4 5 6
INCHES

_______ in.

3.

0 1 2 3 4 5 6 7 8 9 10 11 12 13 14 15
cm

_______ cm

Add or subtract. Show your work in the space below.

4. $36 + 19$	Answer	5. $53 + 27$	Answer	6. $96 - 58$	Answer

Name Date Time

Practice Set 66

Measure the perimeter to the nearest $\frac{1}{2}$ cm.

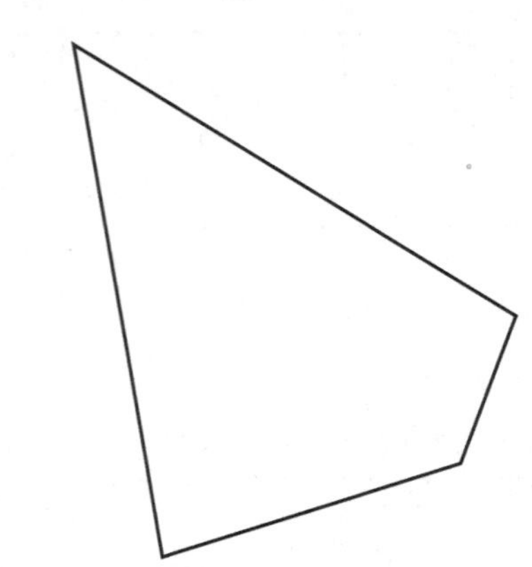

1. **Perimeter:** ________ cm

2. **Perimeter:** ________ cm

Color the part or parts to match the fraction.

3.

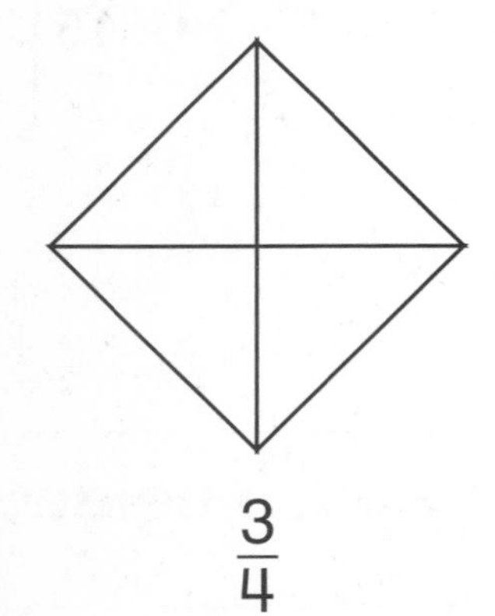

$\frac{3}{4}$

4.

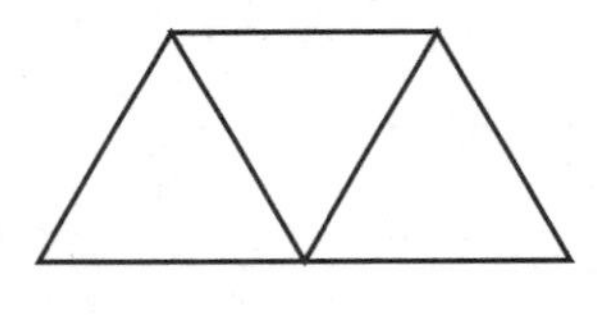

$\frac{1}{3}$

5.

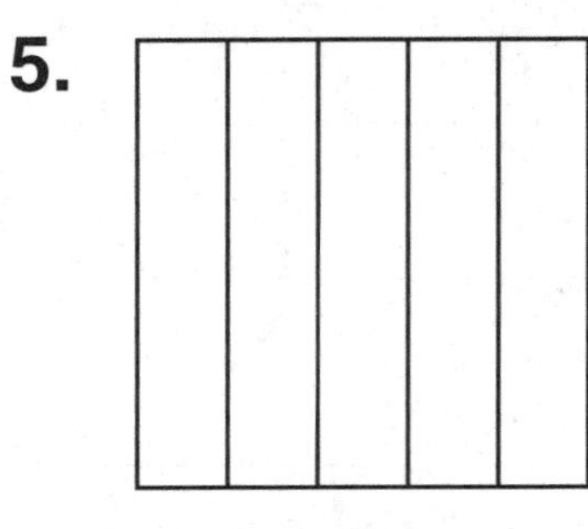

$\frac{3}{5}$

Shade $\frac{1}{2}$ of each picture. Write 2 fractions for each picture.

6.

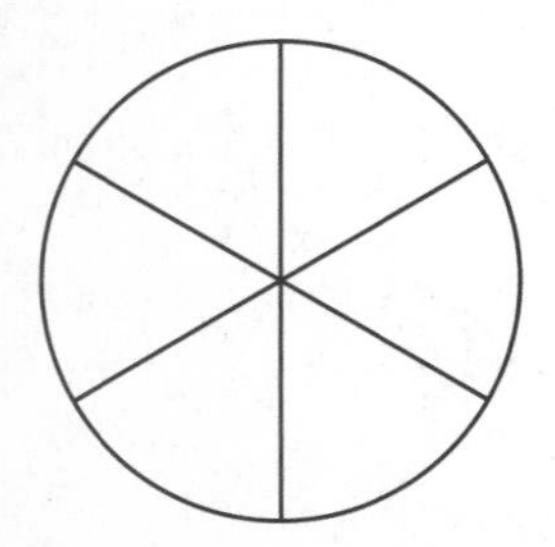

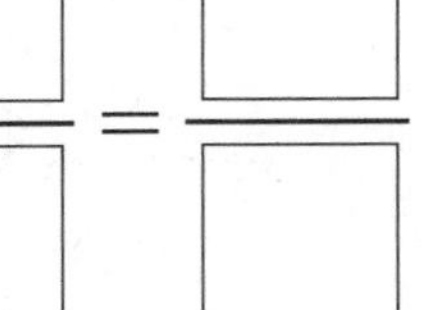

7.

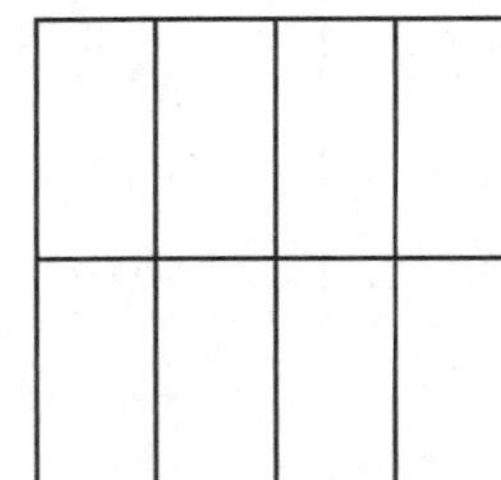

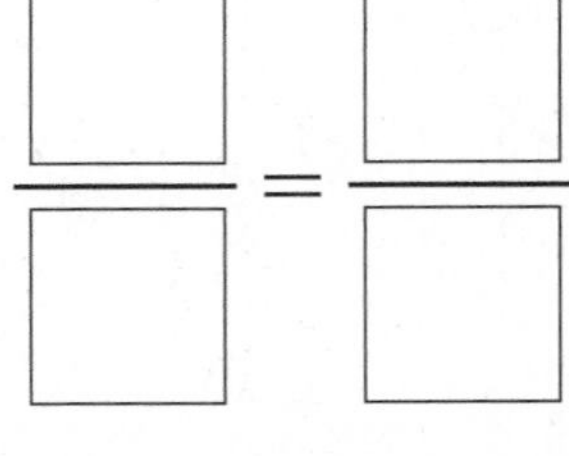

Use with or after Lesson 9•4.

Name ________________ Date ________ Time ________

Practice Set 67

Circle your answers. Which unit would you use to measure...

1. the length of your pencil? inches feet yards

2. the length of a playground? inches feet yards

3. There are 4 white marbles and 1 black marble in a bag. If you pick a marble from the bag without looking, what is the chance that you will pick the black marble?

   unlikely likely certain impossible

4. **Writing/Reasoning** Explain how you found the answer to Problem 3.

   ____

   ____

   ____

5. Solve.

   ______ pennies = \$1.50 ______ nickels = \$1.50

   ______ dimes = \$1.50 ______ quarters = \$1.50

Fill in the answers.

6. How many days are in one week? ______ days

7. How many hours are in one day? ______ hours

Name Date Time

Practice Set 68

Circle the measurement tool you would use.

1. 

2. 

Circle the unit of measure you would use.

3. 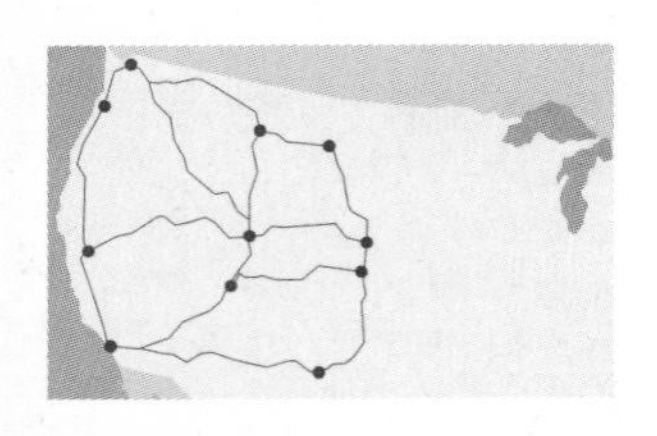

miles feet

4. 

centimeters meters

Use your calculator to find the answers.

5. Enter 39.
Change to 70.
Add or subtract? ______
How much? ______

6. Enter 90.
Change to 24.
Add or subtract? ______
How much? ______

7. Sue entered a number in her calculator.
Then she changed the number to 50
by adding 13. What number did she enter *first*? ______

Name Date Time

Practice Set 69

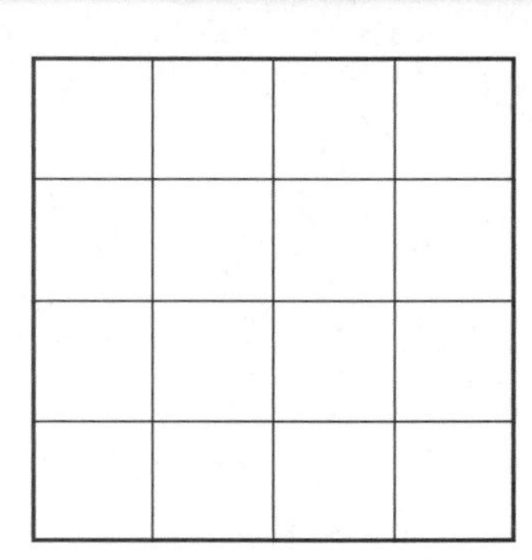

1. Area: ______ sq cm

Perimeter: ______ cm

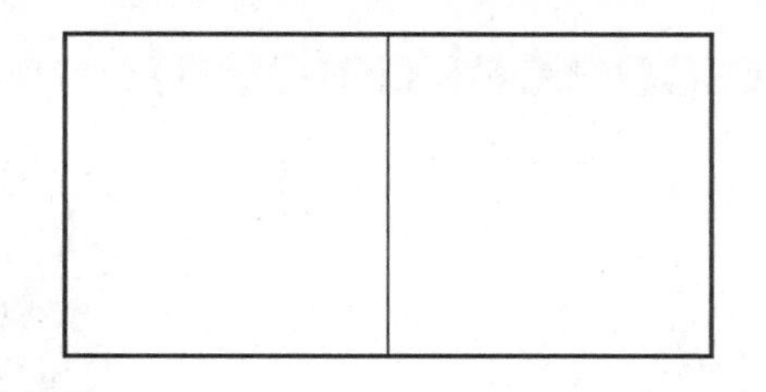

2. Area: ______ sq in.

Perimeter: ______ in.

For each multiplication story

- **draw a picture or diagram**
- **fill in the number model**
- **write the answer**

3. Ruth cut 3 large sandwiches into 3 pieces each. How many pieces does she have in all?

Number model:

____ × ____ = ____

Answer: ____ ________
(unit)

4. Ryan has 3 fishbowls with 4 fish in each bowl. How many fish does he have in all?

Number model:

____ × ____ = ____

Answer: ____ ________
(unit)

Name Date Time

Practice Set 70

Circle the correct amount.

1. 

$\frac{1}{2}$ cup 1 quart

2. 

1 pint $\frac{1}{2}$ gallon

3. 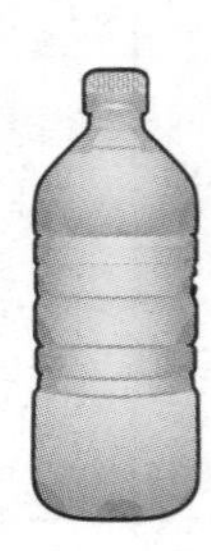

$\frac{1}{2}$ liter 1 milliliter

Write the turn-around addition facts.

Example

4, 11, 7   4 + 7 = 11   7 + 4 = 11

4. 12, 3, 9 ___ + ___ = ___ ___ + ___ = ___

5. 13, 9, 22 ___ + ___ = ___ ___ + ___ = ___

6. 7, 6, 13 ___ + ___ = ___ ___ + ___ = ___

7. 17, 6, 23 ___ + ___ = ___ ___ + ___ = ___

8. 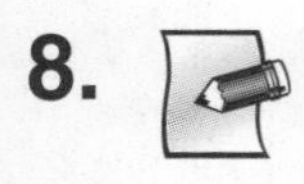 **Writing/Reasoning** Write a multiplication story for this number model: $2 \times 5 = 10$.

____

____

____

Name Date Time

Practice Set 71

1. How much does half a bag weigh?

____

2. How much do 2 bags weigh?

____

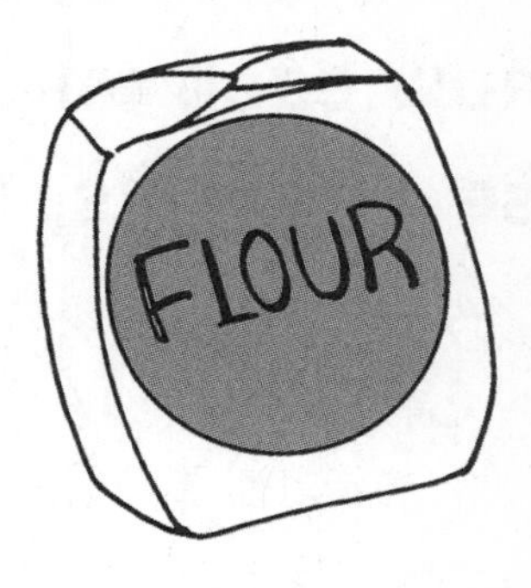

5 kg

Measure each object to the nearest half-centimeter.

Example

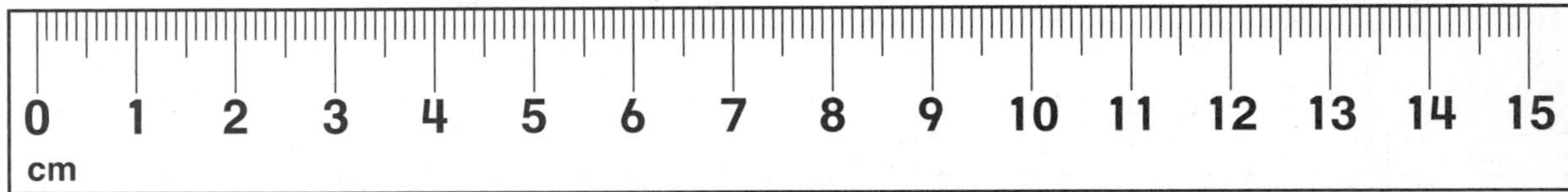

Measurement: about $12\frac{1}{2}$ cm

3.

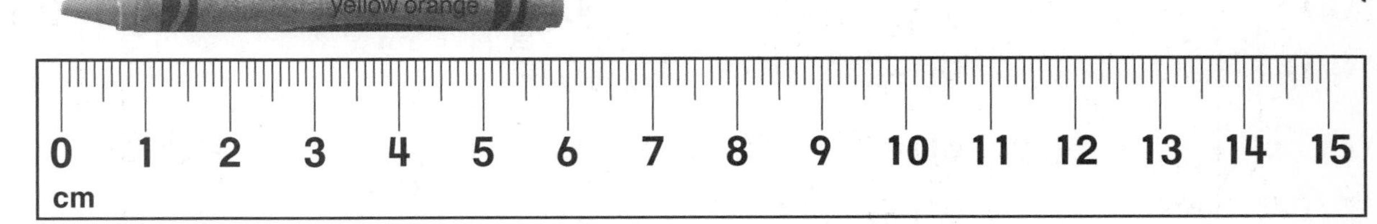

Measurement: about _______________

4.

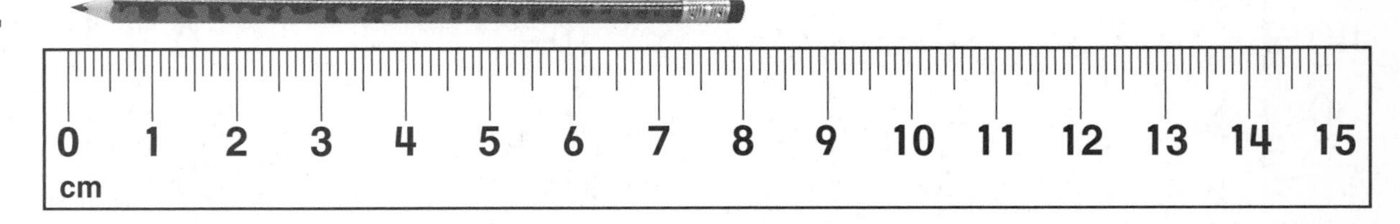

Measurement: about _______________

Name Date Time

Practice Set 72

Show each amount of money two different ways.
Use Ⓠs, Ⓓs, Ⓝs, Ⓟs, and [$1]s.

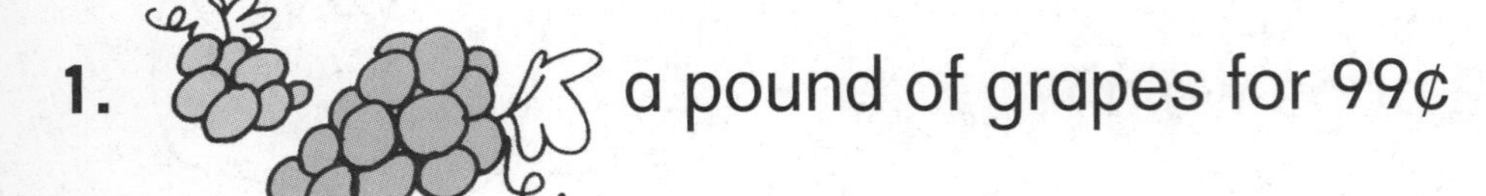

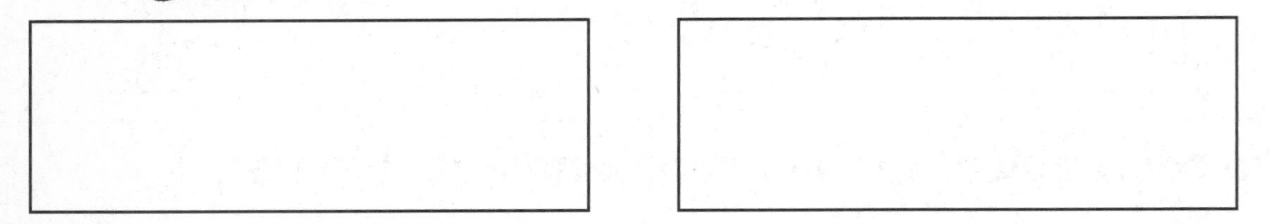

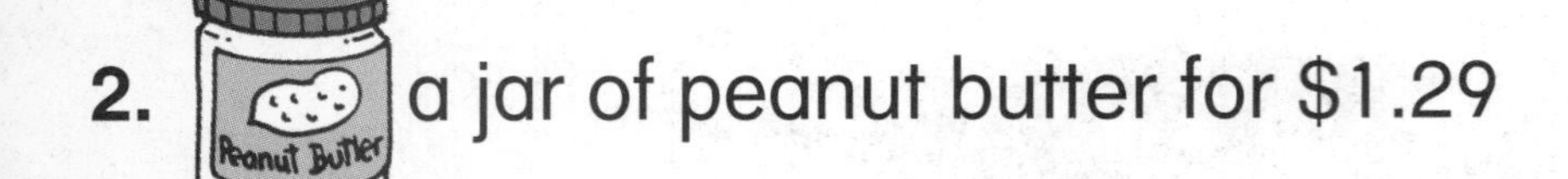

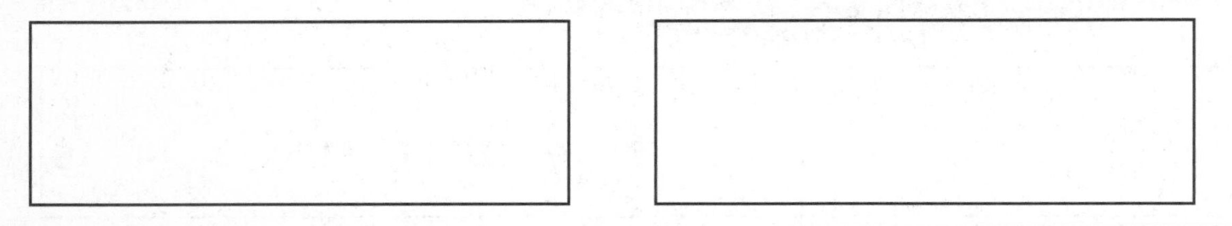

Mark the thermometers to show the temperature.
Circle whether the temperature is warm or cold.

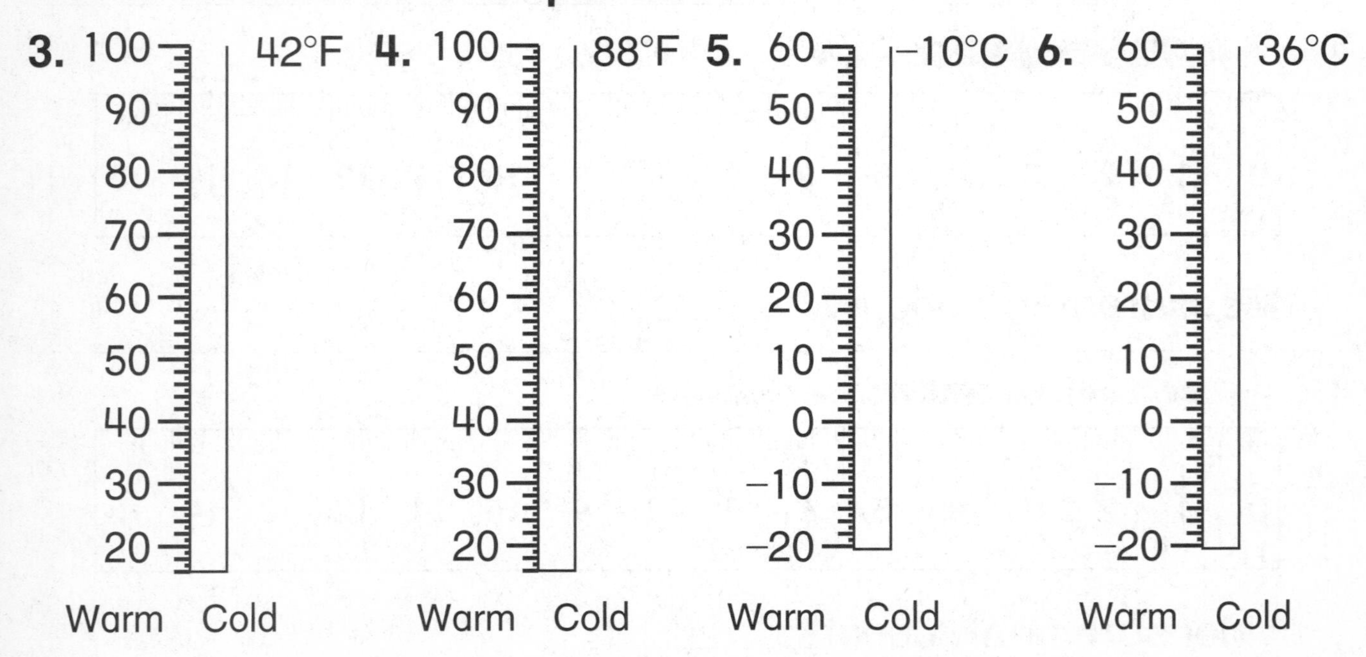

Use with or after Lesson 10•1.

Name Date Time

Practice Set 73

Match equal amounts.

1. $\frac{1}{2}$ dime — $0.01
2. penny — $\frac{1}{10}$ dollar
3. dime — nickel
4. $\frac{1}{4}$ dollar — quarter

Color the part or parts to match the fraction.

5.

$\frac{3}{5}$ are red.

6.

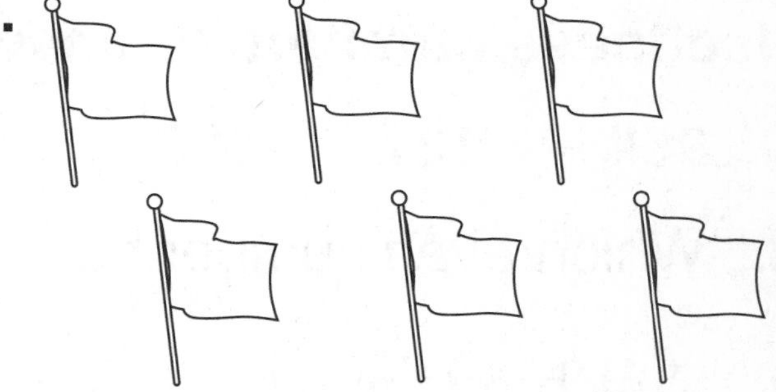

$\frac{1}{2}$ are blue.

Draw these line segments.

7.

$\overline{AB}$
$\overline{BC}$
$\overline{CD}$
$\overline{DE}$
$\overline{EF}$
$\overline{FA}$

B• C• A• D• F• E•

8.

$\overline{AB}$
$\overline{BC}$
$\overline{CA}$
$\overline{AE}$
$\overline{ED}$
$\overline{DC}$

B• A• •C E• •D

Name Date Time

Practice Set 74

Use your calculator. Enter each amount of money. Then write what your calculator shows.

Enter	Display
47¢	0.47
1. $1.32	
2. $0.98	
3. 9¢	

Enter	Display
4. 65¢	
5. $2.50	
6. $0.04	
7. $4.09	

Use these numbers to answer the questions below.

954 127 923 962 106 753

8. Which *even* number has 6 tens? ____________

9. Which *odd* number has 1 hundred? ____________

10. Which *even* numbers have 9 hundreds? ____________

11. Which *odd* number has 5 tens? ____________

12. Which event is **impossible**? Circle the best answer.

A. You will be president of a company someday.

B. The temperature outside will be warmer tomorrow.

C. Thanksgiving will be celebrated by many people.

D. You will be younger next year than you are now.

Name Date Time

Practice Set 75

Buying Groceries

American Cheese $1.49	Saltines 69¢
Wheat Bread 99¢	Hamburger Buns 69¢
Potato Chips 89¢	Mayonnaise $1.99
Ketchup $1.09	Ice Cream $2.49
Watermelon $2.99	Yogurt $2.09
Grape Jelly $1.69	Lunch Meat $1.39

You want to buy the items listed below. *Estimate* the total cost. Then *find* the total cost.

Items	Estimated Cost	Total Cost
1. Potato Chips	about $ ___.___	$ ___.___
Yogurt		$ ___.___
		$ ___.___ Total
2. American Cheese	about $ ___.___	$ ___.___
Grape Jelly		$ ___.___
		$ ___.___ Total
3. Mayonnaise	about $ ___.___	$ ___.___
Watermelon		$ ___.___
		$ ___.___ Total

Name Date Time

Practice Set 76

Use your calculator to solve each problem.

	Items Bought	Total Cost	Amount You Paid With	Amount of Change
1.	Pen: $1.89	$ ___.___	$5.00	$ ___.___
	Notebook: $2.40	$ ___.___		
	Total:	$ ___.___		
2.	Crayons: $1.20	$ ___.___	$10.00	$ ___.___
	Game: $6.85	$ ___.___		
	Total:	$ ___.___		

3. Shade a rectangle that is 15 square units.

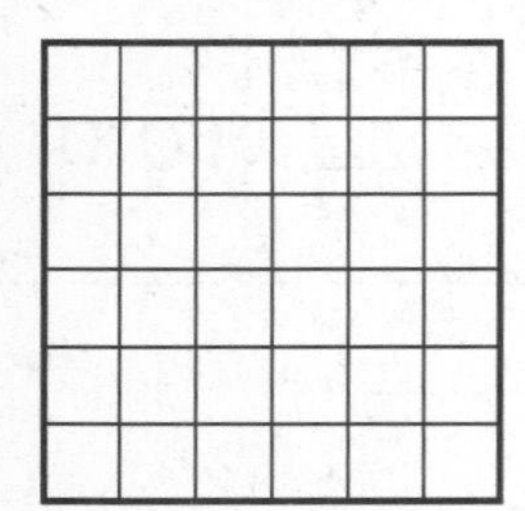

4. Shade a rectangle that is 20 square units.

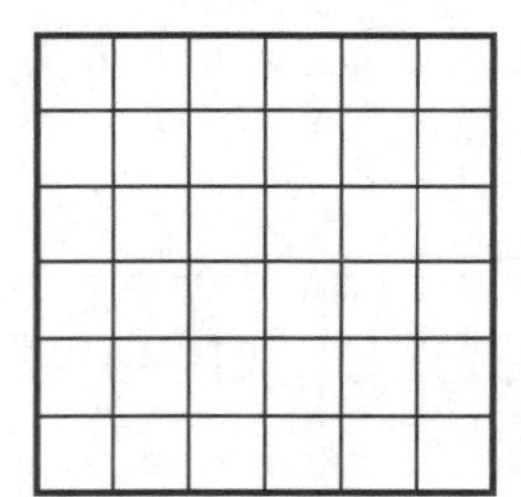

Write the missing number.

5. ______ + 68 = 70

6. 40 = 36 + ______

7. 72 + ______ = 80

8. ______ + 49 = 50

9. 20 = ______ + 11

10. ______ = 85 + 15

Name Date Time

Practice Set 77

Write the amount.

1.

$ ____ . ____

2.

$ ____ . ____

Complete each table.

3.

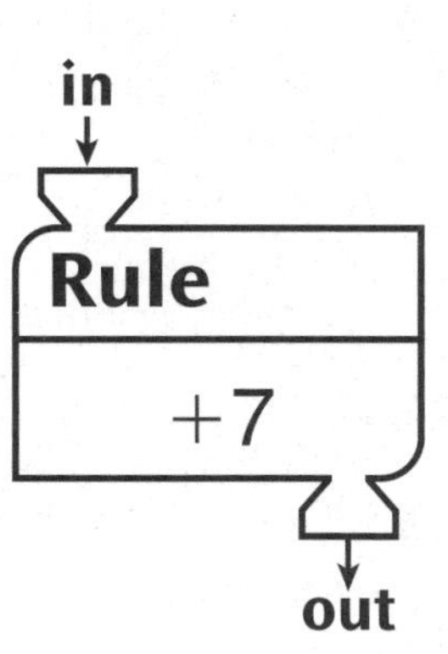

in	out
5	12
	16
	8
	13

4.

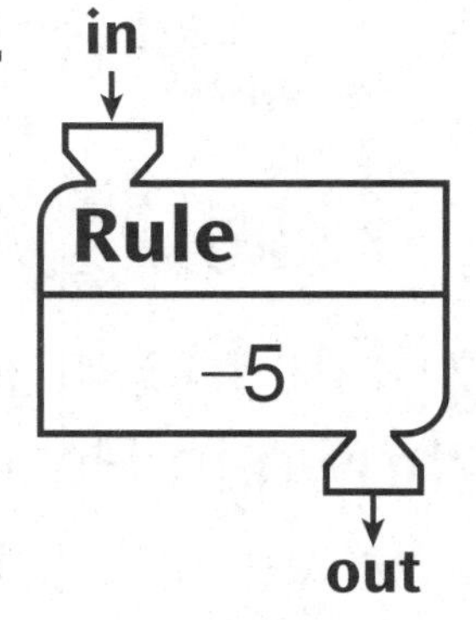

in	out
	10
	13
	17
	45

5.

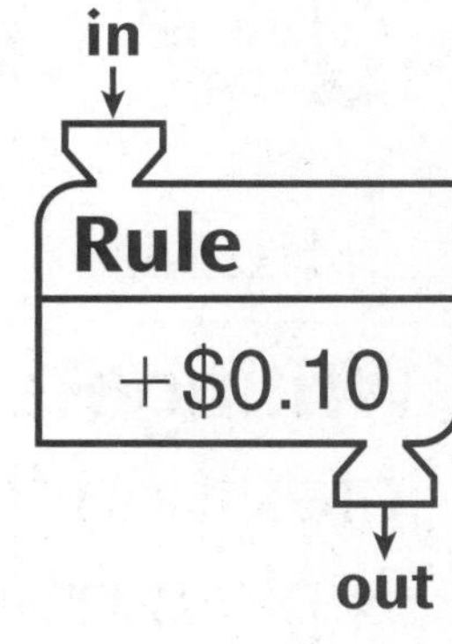

in	out
	$0.20
	$0.35
	$0.18
	$0.12

Name Date Time

Practice Set 78

1. Which lines are parallel? ______

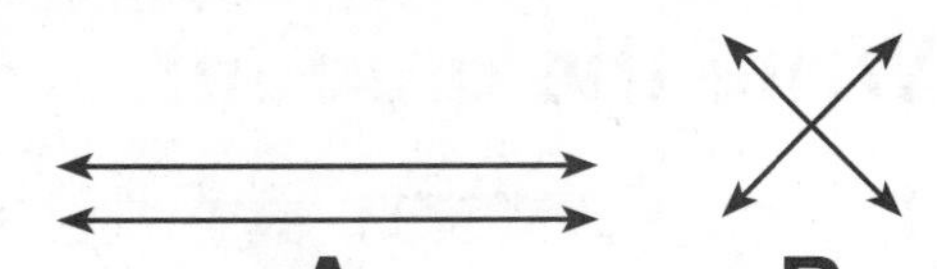

2. Which shape is symmetrical? ______

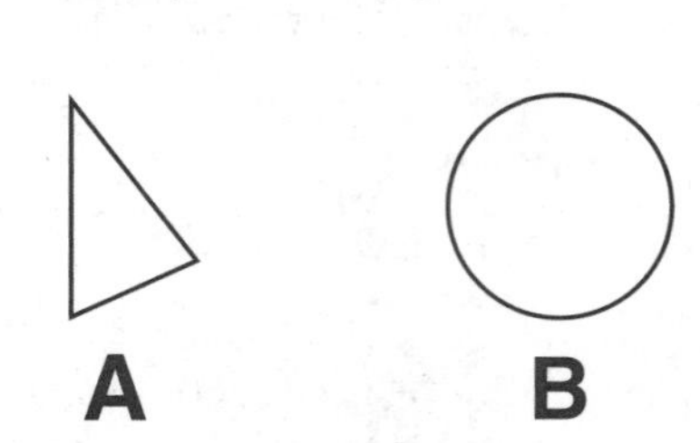

3. Write the time.

___ : ___

4. Draw the clock hands.

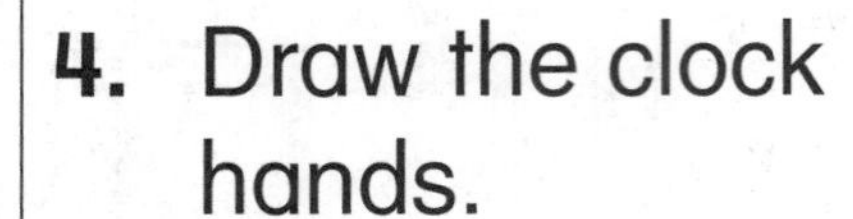

7 : 45

5. Draw the clock hands. Write the time your clock shows.

___ : ___

6. **Writing/Reasoning** Tell the time two hours later than the time shown in Problem 3. Explain how you found your answer.

____

____

7. Show $4.21 using the fewest number of coins and bills. Use Ⓟ, Ⓝ, Ⓓ, Ⓠ, and [$1].

Name Date Time

Practice Set 79

Write the number.

1. four thousand, two hundred ten ______
2. four thousand, twelve ______
3. four hundred one ______
4. ten thousand, six hundred nine ______
5. fourteen thousand, one hundred ______
6. forty thousand, ninety-eight ______

Add or subtract. Show your work in the workspace.

7. 77 − 28 **Answer**	8. 51 + 29 **Answer**	9. 89 + 12 **Answer**
10. 36 − 34 **Answer**	11. 47 + 29 **Answer**	12. 45 + 69 **Answer**

Name Date Time

Practice Set 80

Solve.

1. $17 - (8 + 5) =$ ______

2. $24 - (6 + 9) =$ ______

3. $28 + (6 - 2) =$ ______

4. $13 + (10 - 9) =$ ______

5. $16 + (27 - 17) =$ ______

6. $18 + (16 - 9) =$ ______

7. **Writing/Reasoning** Explain how you solved Problem 6.

____

____

Measure to the nearest half-inch.

8.

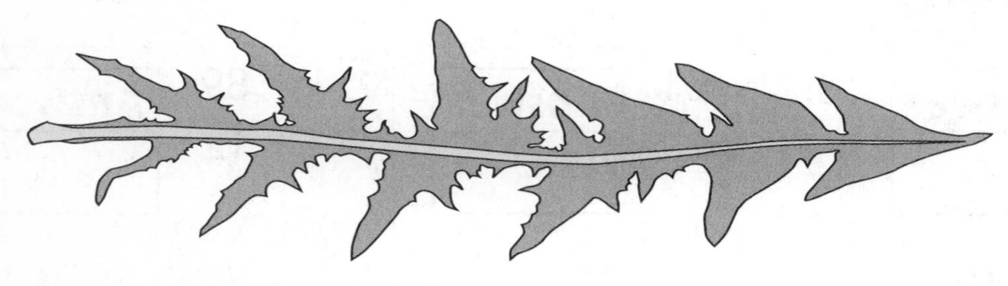

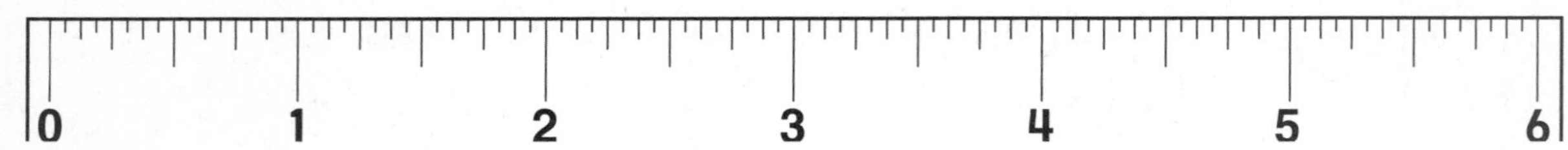

Measurement: about ______________

9.

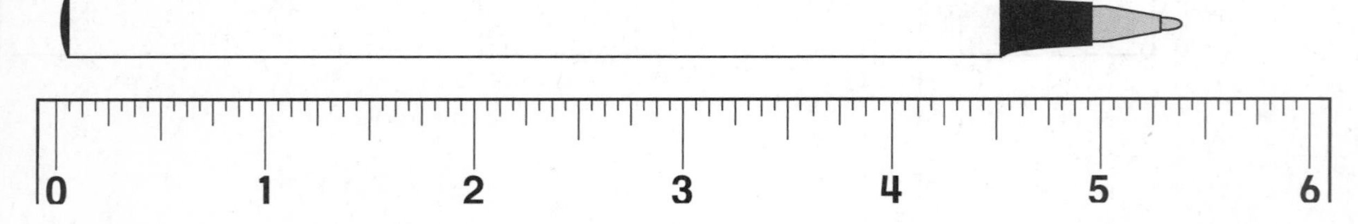

Measurement: about ______________

Name Date Time

Practice Set 81

1. A bag of apples costs $5.21. A box of crackers costs $3.29. How much do they cost together?

   Estimated Cost: ____________

   Total Cost: ____________

2. A box of peaches costs $7.11. A bag of oranges costs $4.69. How much do they cost together?

   Estimated Cost: ____________

   Total Cost: ____________

Complete each table.

3. Rule: $\times 4$

in	out
1	4
	8
	16
	12
5	

4. Rule: $\div 2$

in	out
4	2
	3
	10
12	
	7

5. Rule: $\times 10$

in	out
1	10
	40
3	
	60
5	

Name Date Time

Practice Set 82

1. A bag of apples costs $5.21. A bag of oranges costs $4.69. How much more do the apples cost?

   Estimated Difference: ____________

   Actual Difference: ____________

2. A bag of chips costs $3.29. A box of pretzels costs $2.25. How much less do the pretzels cost?

   Estimated Difference: ____________

   Actual Difference: ____________

Tell what each of the digits stands for. Be careful!

Example 4,703

7 *hundreds*

4 *thousands*

3 *ones*

0 *tens*

3. 2,865

   6 ____________

   8 ____________

   5 ____________

   2 ____________

---

4. 1,639

   9 ____________

   1 ____________

   3 ____________

   6 ____________

5. 3,052

   5 ____________

   0 ____________

   2 ____________

   3 ____________

Name Date Time

Practice Set 83

For each multiplication story ...

- **draw a picture, diagram, or array**
- **fill in the number model**
- **write the answer**

1. Four cars go to the beach. Each car has 5 people inside. How many people are going to the beach?

Number model:

____ × ____ = ____

Answer: ____ ______ (unit)

2. Maria has 9 pairs of shoes in her closet. How many shoes does she have in all?

Number model:

____ × ____ = ____

Answer: ____ ______ (unit)

Write the word that answers each riddle.

cube triangle sphere circle square cylinder

3. I have 3 sides and 3 corners. ____________

4. I am shaped like a ball. ____________

5. I have 4 sides that are all the same length. ____________

6. I am shaped like a square box. ____________

Name Date Time

Practice Set 84

For each division story ...

- **draw a picture, diagram, or array**
- **fill in the number model**
- **write the answer**

1. Peter has 17 flowers.
   He puts 3 flowers in each vase.
   How many vases does he fill?

   _____ ÷ _____ → _____ R _____

   _____ vases have flowers in them.

   _____ flowers are left over.

---

2. Five children share 25 crackers equally.
   How many crackers does each child get?

   _____ ÷ _____ → _____ R _____

   Each child gets _____ crackers.

   _____ crackers are left over.

---

3. Thirty oranges are divided equally among 6 bags.
   How many oranges are put in each bag?

   _____ ÷ _____ → _____ R _____

   _____ oranges are put in each bag.

   _____ oranges are left over.

Name Date Time

Practice Set 85

Draw an array to find the product.

1. $4 \times 3 =$ ______

• • •
• • •
• • •
• • •

2. $2 \times 9 =$ ______

3. $3 \times 7 =$ ______

4. $1 \times 8 =$ ______

5. **Writing/Reasoning** Write a multiplication story for 3×7. Then write a number model.

______

______

______

Solve.

6. On a field trip, the class rode 38 miles on a bus in the morning. In the afternoon, the children rode 27 miles. How many miles did they travel in all?

______

7. How many more miles did the class travel in the morning than in the afternoon?

______

Name Date Time

Practice Set 86

Find the product.

1. $0 \times 7 =$ ____
2. $1 \times 6 =$ ____
3. $109 \times 1 =$ ____
4. $9 \times 0 =$ ____
5. $2 \times 1 =$ ____
6. $10 \times 0 =$ ____
7. $0 \times 50 =$ ____
8. $72 \times 1 =$ ____
9. $123 \times 0 =$ ____

COMPUTATION PRACTICE **Add or subtract. Show your work in the space below.**

10. $87 - 49$ **Answer**	**11.** $16 + 23 + 12$ **Answer**	**12.** $64 - 35$ **Answer**
13. $49 + 53$ **Answer**	**14.** $89 - 44$ **Answer**	**15.** $53 + 28 + 39$ **Answer**

Name Date Time

Practice Set 87

Write the fact family for each Fact Triangle.

Example

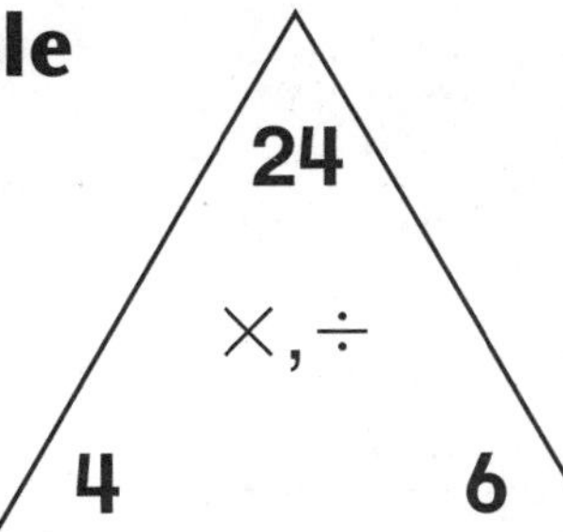

4 × 6 = 24

6 × 4 = 24

24 ÷ 4 = 6

24 ÷ 6 = 4

1.

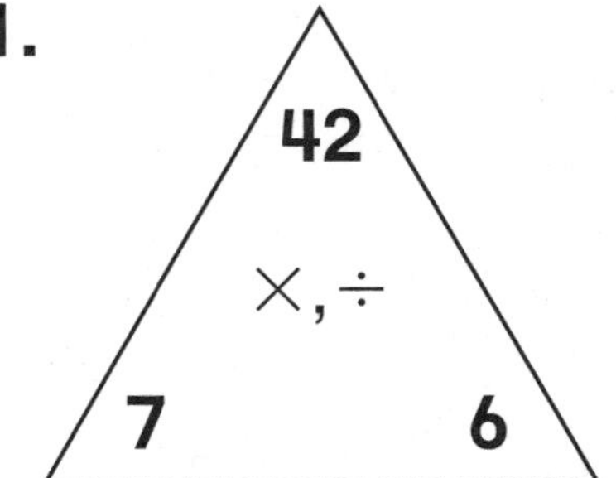

_____ × _____ = _____

_____ × _____ = _____

_____ ÷ _____ = _____

_____ ÷ _____ = _____

2.

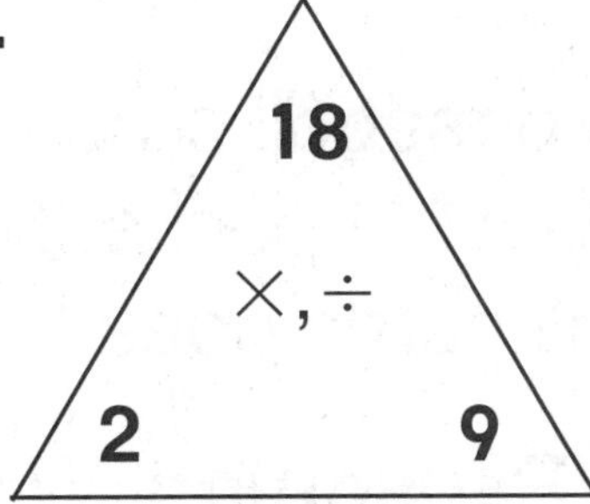

_____ × _____ = _____

_____ × _____ = _____

_____ ÷ _____ = _____

_____ ÷ _____ = _____

For each number below:

Example 9,2 X 6

- **circle** the digit in the hundreds place
- **underline** the digit in the thousands place
- **put an X** over the digit in the tens place
- **draw a square** around the digit in the ones place

3. 5,6 1 7 **4.** 2,0 4 7 **5.** 8,3 9 3

6. 1 0,2 0 8 **7.** 1 2,5 8 1 **8.** 1 6,2 1 7

Name Date Time

Practice Set 88

Complete the number model for each problem.

1. Gia has 20 toy rings. She wants to put an equal number in each of 4 boxes. Draw a picture.

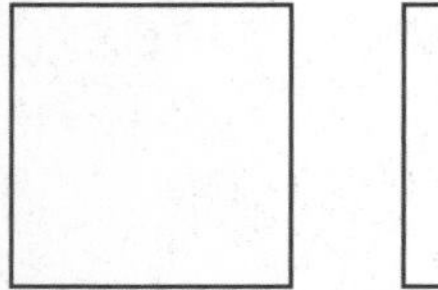 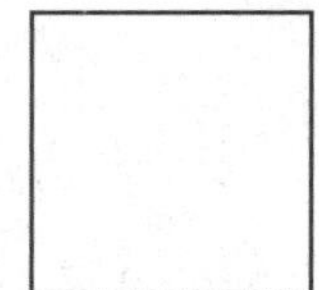 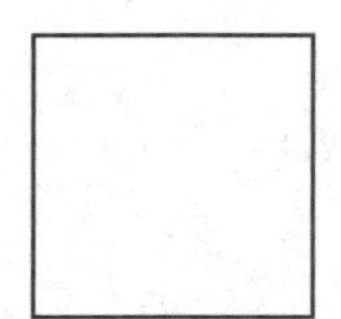 

$20 \div$ ____ = ____

2. There are 5 baseball cards in each pack. John buys 3 packs. How many new baseball cards does he have? ____ × ____ = ____

3. **Writing/Reasoning** Write a multiplication story or a division story. Then write the number model that goes with your story.

____

____

____

4. Fill in the missing numbers.

	2,027	
2,046		
		2,058

Use with or after Lesson 11•9.

Name Date Time

Practice Set 89

1. What month is it? ______________________

2. How many days are in this month? __________ days

3. Write today's date: __________ ________, __________
(month) (day) (year)

Color the part or parts to match the fraction.

4. 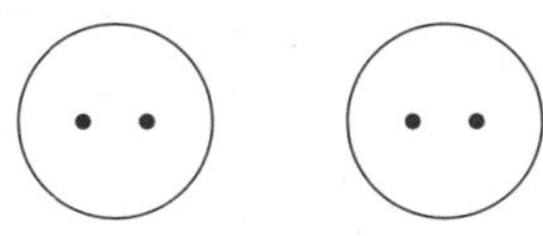
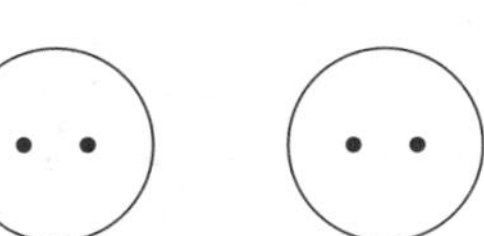
$\frac{3}{4}$

5. 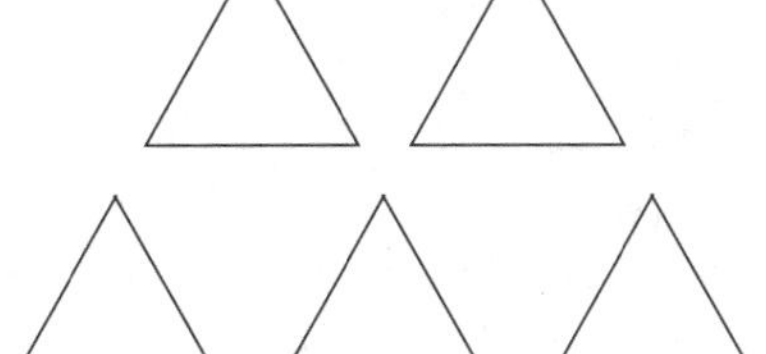
$\frac{1}{2}$

6. 

$\frac{1}{3}$

Draw the other half of the shape.

7. 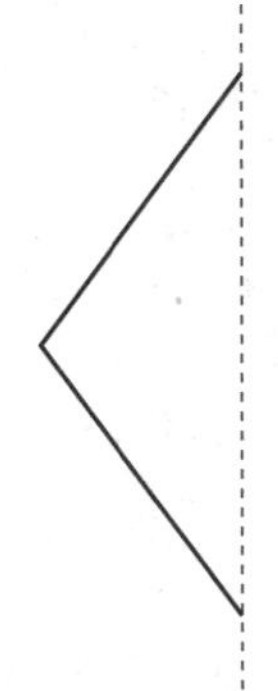

8. 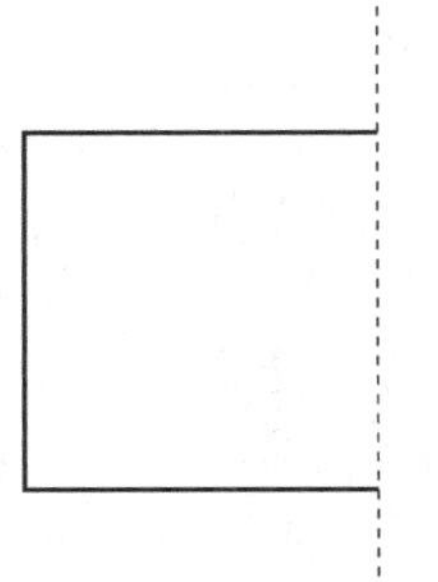

9. 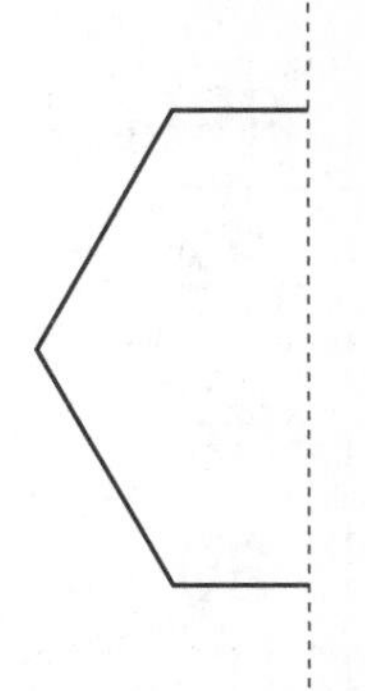

Name Date Time

Practice Set 90

Find the sums.

1. (10 + 4) + 2 = ____

10 + (4 + 2) = ____

2. 13 + 6 = ____

6 + 13 = ____

3. **Writing/Reasoning** Look at Problems 1 and 2. Does it make any difference which two numbers are added first? Explain.

____

____

____

Write the rule. Then fill in the table.

4.

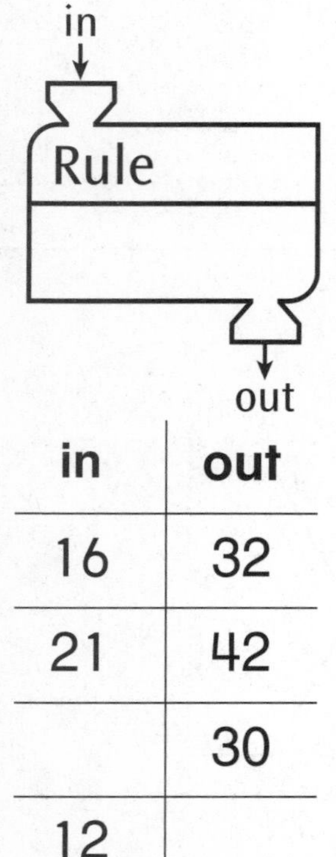

in	out
16	32
21	42
	30
12	
17	

5.

in

Rule

out

in	out
5	15
4	12
6	
	9
	21

6.

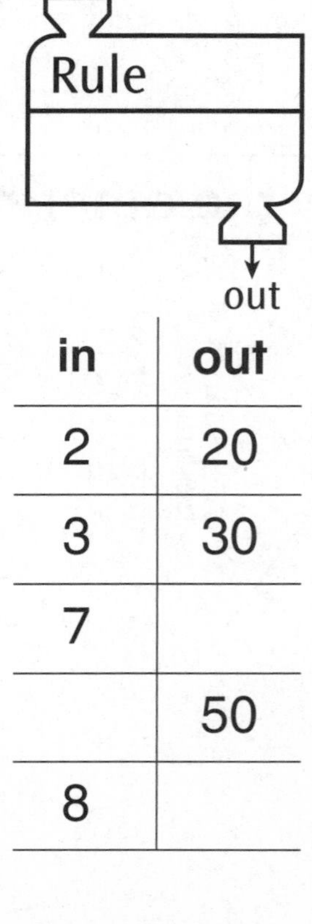

in	out
2	20
3	30
7	
	50
8	

Use with or after Lesson 12•2.

Name Date Time

Practice Set 91

Multiply. Remember to practice your multiplication facts.

1. 9×7

2. 7×6

3. 6×5

4. 8×8

Record the temperature.

5. 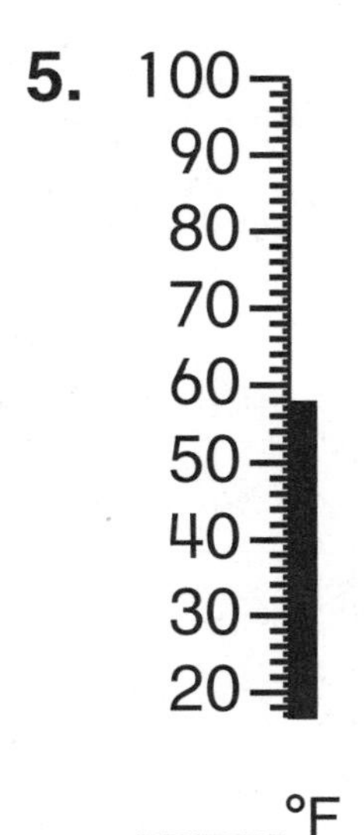

_____°F

6. 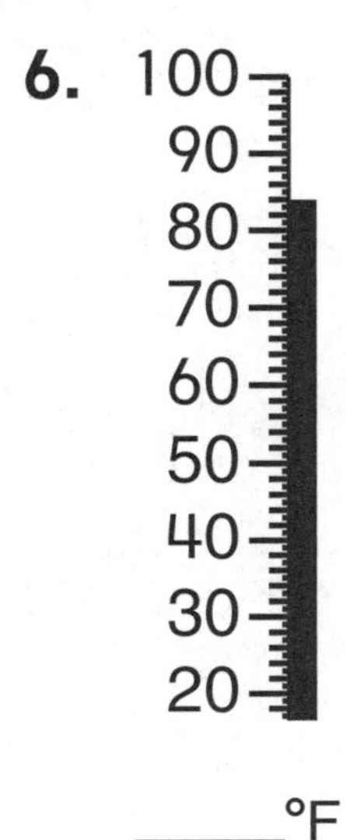

_____°F

7. 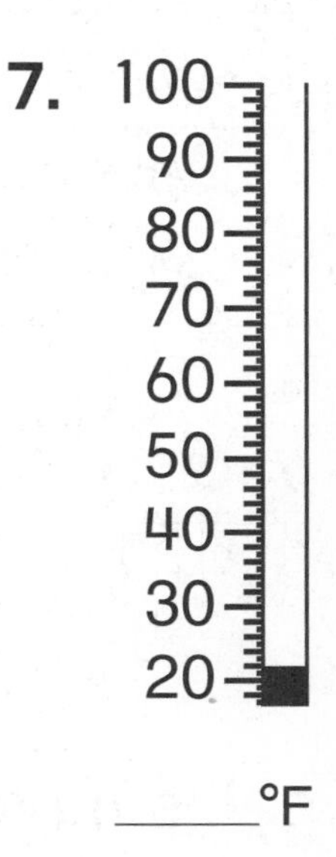

_____°F

Shade the thermometer to show the temperature. Circle whether the temperature is warm or cold.

8. 60, 50, 40, 30, 20, 10, 0, −10, −20 — 6°C

Warm Cold

9. 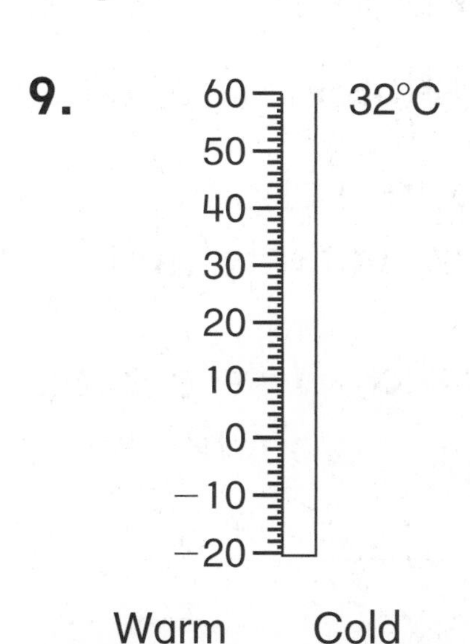

Warm Cold

10. 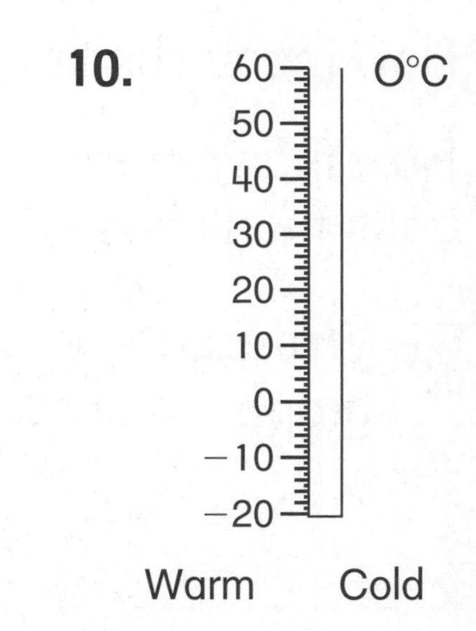

Warm Cold

Name Date Time

Practice Set 92

Mr. Lee's class took a survey of favorite pets. The tally chart shows the results of the survey.

1. Use the tally chart to complete the bar graph.

Childrens' Favorite Pets	
Birds	///
Dogs	~~////~~ ////
Cats	~~////~~ //
Reptiles	//
Fish	/

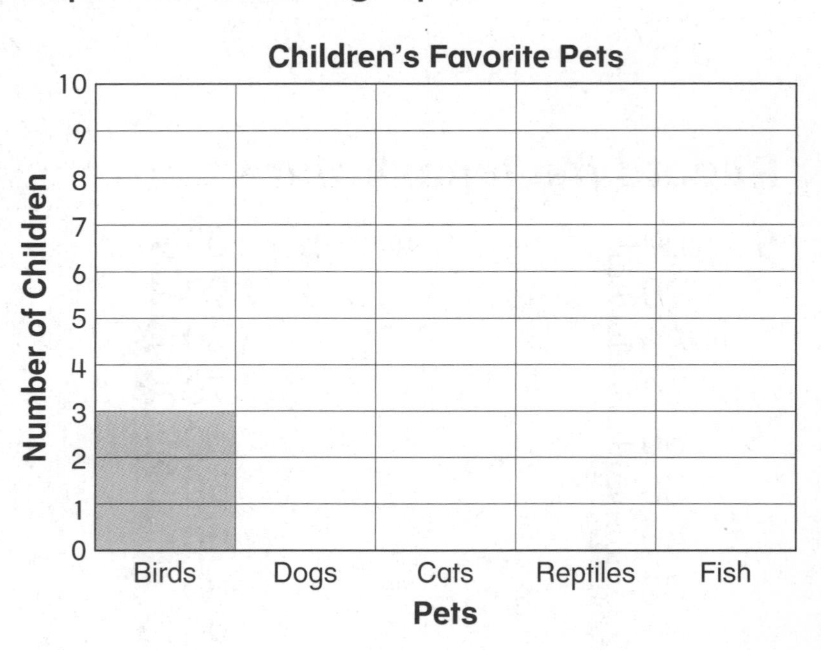

2. Which pet is most popular? ________
   How many children chose this pet? ________

3. Which pet is least popular? ________
   How many children chose this pet? ________

4. How many more children chose dogs than chose cats as their favorite pet? ________

5. **Writing/Reasoning** Write a question about the graph. Tell how you would answer the question.

____

____

Name Date Time

Test Practice 1

Fill in the circle next to your answer.

1. Which number belongs in the box?

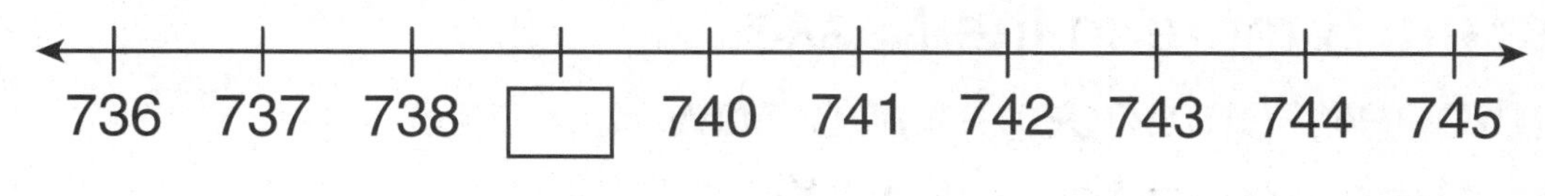

Ⓐ 735 Ⓑ 739 Ⓒ 742 Ⓓ 746

2. What time does the clock show?

Ⓐ 5:30 Ⓑ 6:00
Ⓒ 4:30 Ⓓ 6:30

3. Pete and Rosa played a game. Pete picked a number. Rosa changed Pete's number to a different number. What did Rosa do to change each number?

Ⓐ She added 3 to each number.
Ⓑ She added 4 to each number.
Ⓒ She added 5 to each number.
Ⓓ She added 6 to each number.

RULE: ?

Pete	Rosa
1	6
3	8
6	11

4. Tan spins a game spinner. He writes down the color the arrow lands on.

On which color will Tan **never** be able to land?

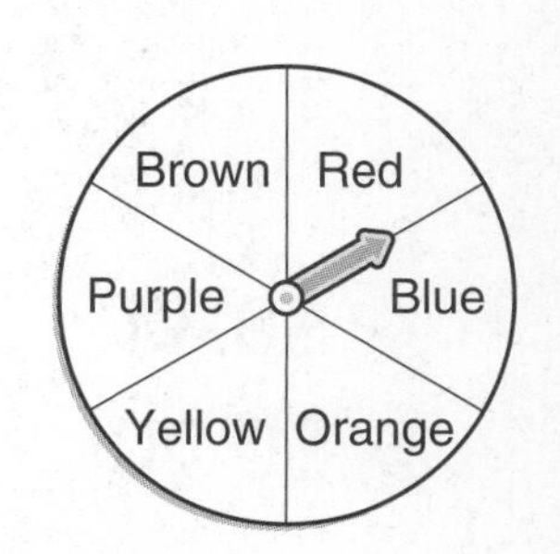

Ⓐ Green Ⓑ Red Ⓒ Blue Ⓓ Purple

Test Practice 1 *continued*

Fill in the circle next to your answer.

5. Hannah put 17 oranges in a basket. Her mom put 5 more in the basket. Which number sentence shows how many oranges are in the basket?

Ⓐ $17 - 5 = 12$ Ⓑ $8 + 9 = 17$

Ⓒ $17 + 5 = 22$ Ⓓ $27 - 10 = 17$

6. Look at Muna's Fact Triangle. Which number did she cover with her thumb?

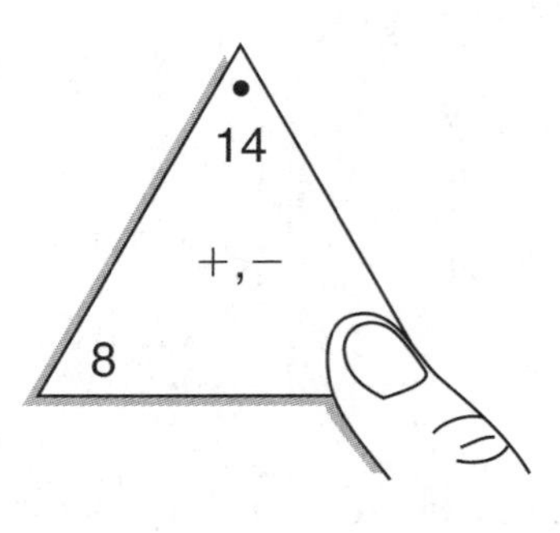

Ⓐ 4 Ⓑ 5

Ⓒ 6 Ⓓ 7

7. Kristy bought pencils with all of the coins below. How much did the pencils cost?

Ⓐ \$0.11 Ⓑ \$0.65

Ⓒ \$0.70 Ⓓ \$1.15

Name Date Time

Test Practice 2

Fill in the circle next to your answer.

1. What is the change in temperature?

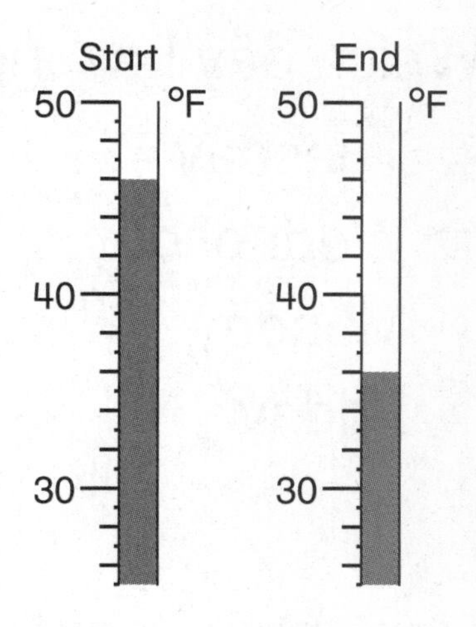

Ⓐ +10

Ⓑ −10

Ⓒ +20

Ⓓ −20

2. Which number model shows a ballpark estimate for 67 + 98?

Ⓐ 60 + 80 = 140 Ⓑ 50 + 90 = 140

Ⓒ 70 + 80 = 150 Ⓓ 70 + 100 = 170

3. Which shape has parallel line segments?

Ⓐ 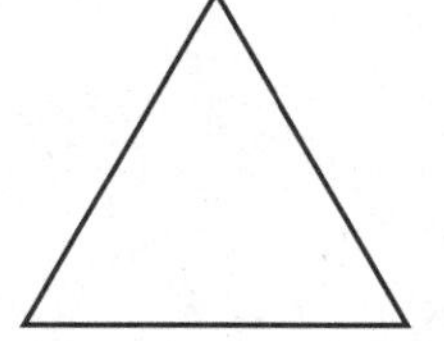 Ⓑ 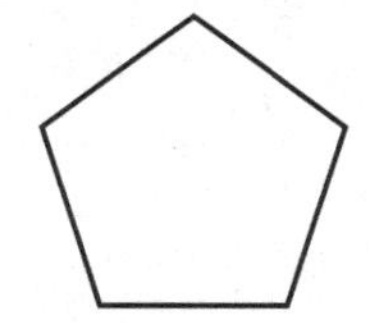 Ⓒ 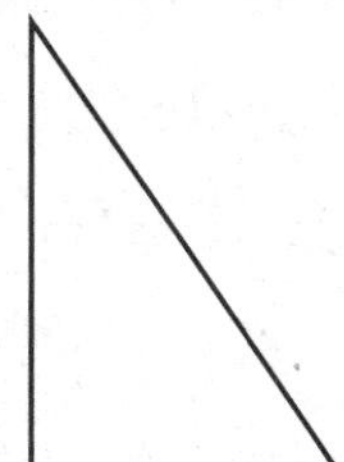 Ⓓ 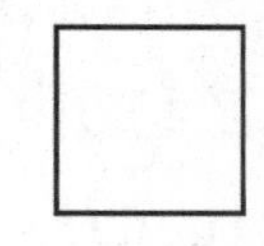

4. Which symbol belongs in the box?

14 □ 23

Ⓐ = Ⓑ <

Ⓒ & Ⓓ >

Name Date Time

Test Practice 2 *continued*

Fill in the circle next to your answer.

5. Which day has the shortest recess time?

Ⓐ Tuesday

Ⓑ Wednesday

Ⓒ Monday

Ⓓ Friday

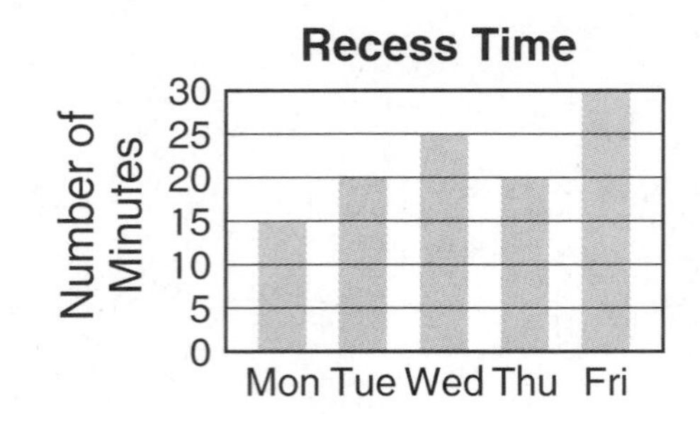

6. How much money?

Ⓐ \$1.06 Ⓑ \$1.11 Ⓒ \$1.16 Ⓓ \$1.36

7. Belen has 18 balloons for her party.
She ties 2 balloons to each chair.
How many chairs are there?

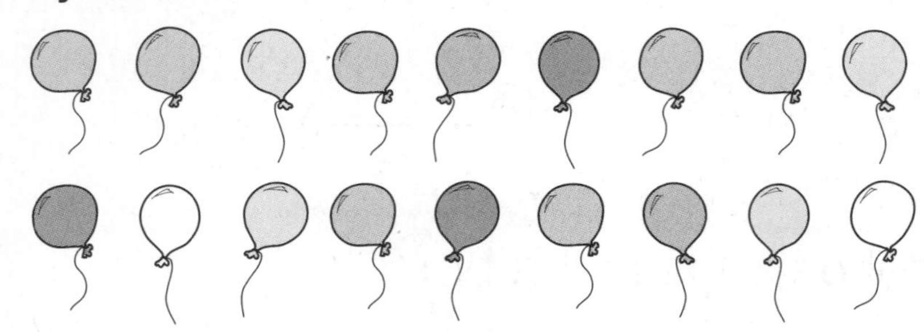

Ⓐ 6 Ⓑ 9 Ⓒ 16 Ⓓ 20

8. Which digit belongs in the box?

$15 - \square = 7$

Ⓐ 2 Ⓑ 7 Ⓒ 8 Ⓓ 22

Name ____________________ Date ________ Time ________

Test Practice 3

Fill in the circle next to your answer.

1. Find the difference.

$$\begin{array}{r} 62 \\ -\ 14 \\ \hline \end{array}$$

Ⓐ 48 Ⓑ 52 Ⓒ 58 Ⓓ 76

2. Find the **median** of the following set of numbers.

18, 35, 21, 14, 11

Ⓐ 35 Ⓑ 21 Ⓒ 18 Ⓓ 14

3. Eight puppies are in a basket. Six are female pups. What fraction of the puppies is female?

Ⓐ $\frac{6}{8}$ Ⓑ $\frac{1}{2}$ Ⓒ $\frac{5}{8}$ Ⓓ $\frac{3}{8}$

4. This chart shows how four baseball teams have played so far this season.

TEAM STANDINGS

Team Name	Wins	Losses
Rams	5	3
Tigers	7	1
Sharks	4	4
Bulls	2	6

Which team is **most likely** to be the first to reach 10 wins?

Ⓐ Rams Ⓑ Tigers Ⓒ Sharks Ⓓ Bulls

Name Date Time

Test Practice 3 *continued*

Fill in the circle next to your answer.

5. Find the **mode** for the following set of numbers.

3, 6, 9, 3, 0, 1, 3, 0

Ⓐ 0 Ⓑ 3 Ⓒ 6 Ⓓ 9

6. Measure the crayon to the nearest $\frac{1}{2}$ inch.

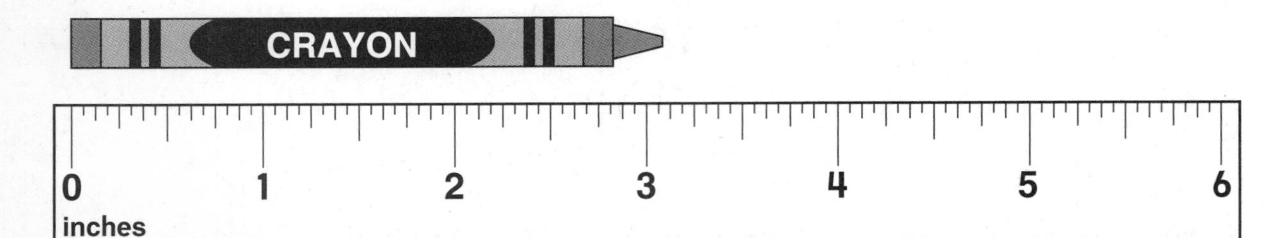

Ⓐ 3 in. Ⓑ $3\frac{1}{2}$ in. Ⓒ $4\frac{1}{2}$ in. Ⓓ 4 in.

7. Find the perimeter of the rhombus.

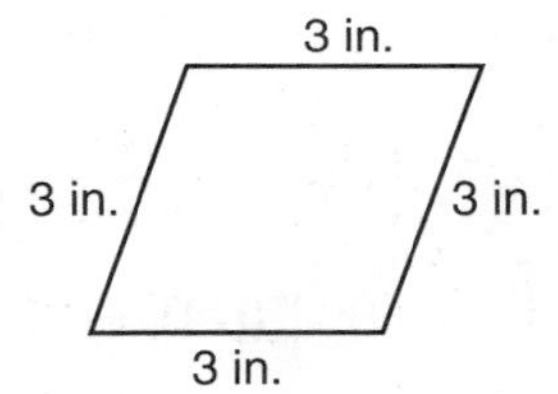

Ⓐ 12 inches Ⓑ 9 inches
Ⓒ 6 inches Ⓓ 3 inches

8. What is the missing number in the rule chart?

Rule: 1 cup = 8 oz	
cup	oz
2	16
3	24
5	?

Ⓐ 20 Ⓑ 30 Ⓒ 32 Ⓓ 40

Name Date Time

Test Practice 

Fill in the circle next to your answer.

1. How many hours are in 2 days?

Ⓐ 12 Ⓑ 24 Ⓒ 36 Ⓓ 48

2. What does the 2 in 2,785 stand for?

Ⓐ 2 ones Ⓑ 2 tens

Ⓒ 2 hundreds Ⓓ 2 thousands

3. What fraction of the dots is circled?

Ⓐ $\frac{8}{12}$ or $\frac{2}{3}$ Ⓑ $\frac{4}{12}$ or $\frac{1}{3}$

Ⓒ $\frac{2}{12}$ or $\frac{1}{6}$ Ⓓ $\frac{1}{12}$ or $\frac{1}{6}$

4. A sign at the Kennedy Space Center says that it opens at 9:00 A.M. What can Michael use to find out how long he must wait until he can enter?

Ⓐ 

Ⓑ 

Ⓒ 

Ⓓ 

Name Date Time

Test Practice 4 *continued*

Fill in the circle next to your answer.

5. What time is shown on the clock?

Ⓐ 4:25 Ⓑ 5:25 Ⓒ 4:20 Ⓓ 5:20

6. Davis has 4 trays for his plants. He puts 8 plants in each tray. How many plants does Davis have in all?

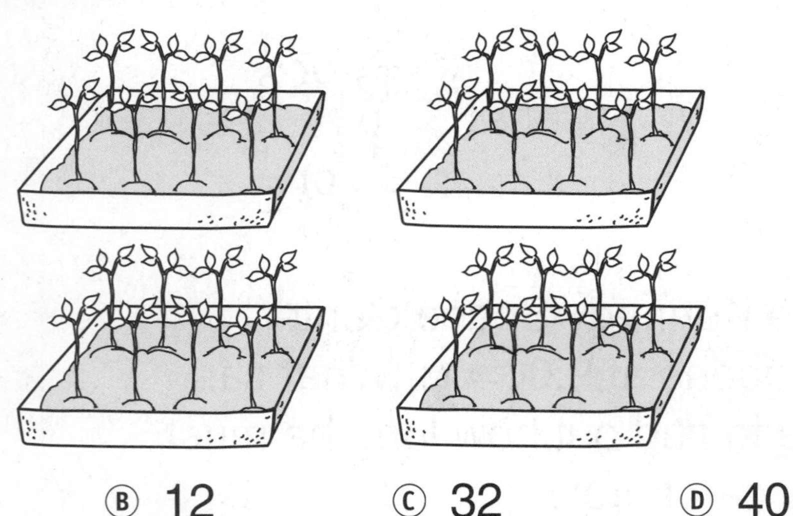

Ⓐ 8 Ⓑ 12 Ⓒ 32 Ⓓ 40

7. Mr. Wong has 15 cans. He wants to put an equal number of cans on each of 3 shelves. How many cans will he put on each shelf?

Ⓐ 3 Ⓑ 5 Ⓒ 15 Ⓓ 18